AutoBioDiversity

AutoBioDiversity

True Stories from *ZYZZYVA*

edited by Howard Junker

HEYDAY BOOKS
BERKELEY, CA
2005

Cover Image: Christopher Brown, *Fourth Gate,* 2001, 46 x 36 1/4 inches, color intaglio print, courtesy: the artist and Paulson Press, Berkeley, part of an edition of 35 donated by the artist and the press to be sold for the benefit of *ZYZZYVA.*

Publication supported in part by grants from the William and Flora Hewlett Foundation and the Zellerbach Family Foundation.

Orders, inquiries, and correspondence should be addressed to:
Heyday Books
P.O. Box 9145
Berkeley, CA 94709
www.heydaybooks.com

Library of Congress Cataloging-in-Publication Data
AutoBioDiversity : true stories from *ZYZZYVA* / edited by Howard Junker.
p. cm.
ISBN 1-59714-007-4 (pbk. : alk. paper)
1. American essays—Pacific Coast (U.S.) 2. Pacific Coast (U.S.)—Intellectual life.
3. Pacific Coast (U.S.)—Social life and customs. I. Junker, Howard. II. *ZYZZYVA.*
PS569.A97 2005
814.54080979—dc22
2004029106

Printed in the United States of America
10 9 8 7 6 5 4 3 2 1

FOREWORD

Howard Junker

> *The universe is a big and lonely place. We
> can do with all the neighbors we can get.*
> Bill Bryson
> *A Short History of
> Nearly Everything*

The best thing about editing a literary magazine in San Francisco for the past 20 years has been the simple pleasure of the job itself—the daily routine, the constant struggle. Surviving so long, in such a fragile vehicle, has amazed and delighted me.

And then there's the thrill of finding needles in the haystack of the slush pile! And the satisfaction of helping at least a few writers along their difficult way! And the relief when the issue goes off to the printer! And the manic epiphany when I realize that this is the best issue I've ever done, possibly the best issue of a literary magazine ever produced anywhere.

So much for boilerplate. The really best thing has been the chance to use writers to articulate my own inexpressible needs and desires. If writers use me to take them public, I appropriate them. I use them to flesh out my own dreams. To fill the void in my own imagination. To tell, in a wonderfully camouflaged way, my own story.

To celebrate *ZYZZYVA*'s 15th anniversary, I wrote-and-compiled a "corporate autobiography." I combined a few chapters of naked memoir with a few basic documents—our beloved rejection letter; our down-to-earth writer's contract—and a few bits of how-we-do-it-around-here. I thought this collage made a terrific statement about the intertwining—and over-determination—of the institutional and the the professional and the personal. Many (otherwise loyal) readers couldn't wait til things got back to normal.

At first, out of a sense of decency, I thought I'd skip a 20th-anniversary anthology. We've already done four: after the first five years, a collection of essays; after ten, Our Greatest Hits *and* a collection of facsimiles of writers' notebooks, a regular feature in that era; after 15 years, the corporate autobiography, plus a gathering of accounts of how-I-became-a-writer, a regular feature in *that* era.

But it's hard to pursue business as usual all the time. In fact, the anthologizing process—the intense review and self-scrutiny—tends in and of itself to revive the spirits (and prevent burn-out). It's easy to forget, in the trenches, the good stuff that somehow got published long ago. It's also nice to come upon a pile of material on which the heavy lifting has already been done, and all you have to do is say, I'll take this one and this one and that one.

It may be, however, that it's a little late in the game to be coming to market with a basketful of confessions and memoirs. A lot of our hipper critics have been declaring lately that the entire genre of autobiography is dead—too much self-indulgent gushing.

And from a different perspective, the avant-gardistas have long been trashing the very notions of "self" and "narrative" and "coherence."

I find myself dangling over—or is it among?—these abysses.

In the end, I'm on the side of innocence, and this anthology assumes, across a broad range of personal statement, that the narrator stands behind the prose the way the government stands behind the greenback. (A couple of pieces, in fact, although totally truthful, were original billed as "fiction"; can you spot them?)

Most of the writers are middle-aged, because it usually takes a while to get a life, and the adventures they describe tend not to be exotic.

Phil Levine is the only Celebrated Writer on hand; the Usual Suspects are conspicuously absent. This is in keeping with my essential commitment: to mine the slush pile in order to give new voices a chance. On the other hand, the Queen Mother does make a cameo appearance.

Most of the pieces are recent, although a few have been taken from the last century, to lend historical perspective.

The main story my alter egos tell is simple: Life is a struggle. And the moral is, as the title suggests: It takes all kinds.

In conclusion, I would like to thank the redoubtable publisher and advocate Malcolm Margolin, himself celebrating 30 years of Heyday Books, for insisting that we, too, might do a little celebrating.

Thanks to all of you who helped us get here.

CONTENTS

The Geography of Reading

Bill Mohr

I read *The Kid Comes Back* by John R. Tunis after not being picked for any of the Little League teams in Norfolk, Virginia, for the third consecutive year.

I read *The Brothers Karamazov, The Sound and the Fury,* and *Under the Volcano* in Sandi's student-housing apartment in Boulder. She had a tiny bit of earth beside her front door in which she grew herbs. One day she came back from running experiments about the differences in time required to comprehend metaphorical and literal sentences and found that someone had taken all her mint leaves.

The Bat Poet was a birthday gift from the first woman I ever slept with. Later, another woman gave it to me as a surprise present, and she didn't understand why I couldn't accept it.

I read *The Lord of the Rings* on a farm in Leeuwarden, Holland. My relatives were amused when I tried to pronounce my mother's maiden name, Van Schelven, though it wasn't their name. One Saturday, we drove to a rest home in Friesland to visit the last woman with that name. We couldn't talk to each other, but she seemed very glad to see me.

I read Camus' *The Plague* at lunch and on 15-minute breaks from running a blueprint machine at an architect's.

Halfway up the hill of Westerly Terrace, I read John Donne to Anna in her tiny apartment and realized his poems are for lovers of poetry and not poetry for lovers.

I never finished *Tess of the D'Urbervilles, Sons and Lovers,* or *Moby-Dick*. I've never even started anything by Dickens, Flaubert, Stendhal, or Kafka. But maybe I'll get around to them, as I finally did to *The House of Seven Gables,* leaning back in a simple black chair with my feet up on an old wooden desk I bought for $75.

On top of the Department of Social Services office on Pico Blvd., I read *One Hundred Years of Solitude* on my lunch break. All morning I would talk to people who didn't have enough money to see a doctor. Sometimes, when I felt too desolate to read, I'd walk past liquidambar trees to railroad tracks and then alongside pink-and-white oleanders toward an abandoned tree house.

I read *Going After Cacciato* while working at INTELLECTUALS AND LIARS Bookstore.

I read Lucian's *My Dream* to my students at Phoenix High School in 1993.

There was lightning in the summer night in Bismarck, North Dakota, as I read David Antin's *Talking at the Boundaries.*

I read *Sense and Sensibility* a couple of years before my wife and I separated.

While sick with a cold for four days, I read the first half of *Tom Jones.* I slept and read and slept and read. I was disappointed when I got well and had to go back to work. It was an 18th-century rhythm, and I've never danced to it since.

Not all of the books I've read have been published. My favorite manuscript is *Greenglass* by Charles Baxter, which I read on a park bench on a cloudy day in Palisades Park.

The first time in six years I had two weeks off, I stretched out on the scraggly lawn of an unrented house next to my apartment with a bag of pistachios and *The Rise of Silas Lapham.*

No matter what I'm reading, or where, I keep hoping to find a perfect sentence, something brief, entire, and completely spinning within the stalk of itself, with no reference to any character, so it doesn't need a context, but rolls like a marble in a vacuum tube, a marble as beautiful as those clear ones with green and blue ribbons inside I used to knuckle at the edge of a circle in dirt. □

Bill Mohr completed his Ph.D. at UC San Diego last year and now teaches at Nassau Community College on Long Island. His most recent book is Hidden Proofs *(Bombshelter Press, Los Angeles).*

Guide to Mental Hospitals, Facilities, Halfway Houses & Board-and-Care Homes of the Bay Area

Craig Diaz

MENTAL HOSPITALS

S.F. General: I have suffered immensely at SFGH. My worst West Coast bad drug reaction was there. Every time you go through downstairs Psych Emergency Services you go thru pain. Why then have I admitted myself to SFGH so many times?? Why, I'm an idiot without pride. Typically, I have been admitted and thrived on the Hispanic Ward despite the fact that I speak little Spanish. I tried to sue Dr. Blake for my bad reactions, getting ward notes, but got nowhere in a lawsuit. I did get a nice note from Mayor Feinstein.

St. Francis: The main thing is they serve you huge amounts of food. St. Francis has no religious connection in the reality of the place. When we had outings around the Nob Hill neighborhood, we had some pretty good fun. Maybe I should of been more dogmatic and forceful after bad drug reactions. The staff might have listened. I have been committed to St. Francis three or four times.

French Hospital: Major factors were: the insistence by staff that they were an average hospital, that they had free ice cream and other snacks, that I got along well with staff. I got high on coffee at a nearby Winchell's and was thus depressed at other times. I wrote a lot at French. Had no real bad drug reactions. Psychiatric Ward now closed. I was committed to French twice. I once threw chairs at the staff.

Langley Porter: Supposedly, Langley Porter helped make some of the psychiatric drugs I have suffered so on. I don't care—Langley Porter is a good hospital with good programs. A big question was Langley Porter part of UC? The staff was always wondering. Food was good. I've stayed at Langley Porter three times. Once, with no indication, they just suddenly let me go. Suddenly I was free.

McCuhleys: I had a short time at McCuhleys and it was all right. I ate all their snacks cause I was hungry.

St. Lukes: I should have been treated better as an Episcopalian in an Episcopal hospital. Overall, St. Lukes wasn't too good or bad. Never as good as "average" French. I could never get a feel of the unit. When threatened with a Prolixen shot, I escaped.

El Dorado: I had migraine headaches at El Dorado. I had fun playing table tennis. I went on a lot of outings. I was patient-government president. I drank a lot of coffee. I started my writing career at El Dorado in San Jose. El Dorado was closed a long time ago for having a "rape-riot." I was there a long time. They had a lot of low-level patients at El Dorado who were a mess.

FACILITIES

Crestwood San Jose: I had a lot of broad heterosexual experience at Crestwood San Jose. I knew and had good times with several girls. Beatrice came into my room like a colt, and on coffee we had good times.

I volunteered for staff every day. This was not smart.

My I.D. was stolen by staff.

I exercised a lot. I ran and ran.

We got a free pack of cigarettes every day.

I escaped from Crestwood San Jose.

Crestwood Vallejo: They had tough kids at Crestwood Vallejo. A lot of sexuality, mostly heterosexuality. A good program. Long waits for food. Only two people to a room. I spit out my meds all but twice, had bad, bad, bad drug reactions those two times. I exercised a lot. We patients were eating a lot of cake. I dug a hole under the fence and escaped and returned many times.

Canyon Manor: 30 miles north of San Francisco, an excellent facility. I was into coffee and sometimes the coffee beat me. It was a bit of a degenerate place. I had much fun, wrote Books I lost. I wrote "At Canyon Manor." The soul of "At Canyon Manor" was destroyed by a Haitian girl on the staff who took its varnish right out of it, destroying its sense. I finally beat the point system at Canyon Manor, getting the highest point total. I had many girlfriends. I escaped over and over again to San Rafael or Darwin's, uh, pad in San Francisco. I was president of the patient government. I had several scary fights. Canyon Manor had dances twice a week. I was stupid to escape and escape. They had MTV at Canyon Manor. MTV is very entertaining. I had lots of good times at Canyon Manor.

Napa: I wanted to come to Napa as a journalist. It didn't work out that way. I'm at Unit Q-5. I have friends. The food is good. I like my psychiatrist.

I have a lot of boring depressed days. I lost a Book 118 pages into it.

I have been scared of violence half the time I've been here. I exercise a lot.

I have a lot of friends. Rick introduced me to David, who lives on the outside and has helped me a lot. A virgin, looks like Romanov, is romantic, I like him a lot.

Coffee and cigarettes are important here at Q-5.

The projects are a dismal side-unit. I wanted to stay in them under the blankets for three days to quit smoking.

I tried LSD for the first or tenth time at Napa.

I like going to the courtyard outside. We rarely go outside.

A girl here just went thru hell. She was hit by a chair. I brought her a gift. It was a terrible tragedy.

They don't let you cheek meds.

I hate Napa.

HALFWAY HOUSES

West Side Lodge: Oh, I was there about, oh, '83 or '84 or '85. It was a very nice place. We talked and talked and talked. Group therapy, fun outings. Horrible migraine headaches. They are nobody's fault. Trips to Marin and the East Bay. Camaraderie of young men and women. Well run by Ernie X, unless he's dead; he had a heart condition.

Much more freedom than at hospitals or facilities. There's the right to refuse drugs.

Robertson House: I was at Mandala House twice. Its name was changed to Robertson House since I left. Clients fixed the meals. We had many outings. We had fun. The house was so nice—bit of a hippie pad—but we had responsibilities. I copied page after page in some writing project. We had to go to day-treatment programs every day. The staff were good and you could refuse drugs. We ate well. I had a beer on Superbowl Sunday.

BOARD-AND-CARE HOMES

Chateau Agape: Famous Chateau Agape is on Guerrero Street in the Mission District. Famous leader is Leroy Looper, who is always—he loves publicity—on TV and in the newspapers extolling Chateau Agape as the best board-and-care house in San Francisco. It is a good living situation. I had many good friends. I went to City College while at Chateau Agape. Nice intelligent young people. We had esprit de corps and enthusiasm for the Chateau. The food was O.K. I wrote a lot. The Looper kids were neat. The Loopers were a mixed racial family.

My opinion, overall, of the places I have lived in the Bay Area is not as good as it sounds. Why did they build an extensive young peoples' homeless social state?? □

This was Craig L. Diaz's first paid appearance in print.

TUMBLING INTO WRITING

Susan Parker

I was not a writer before my husband, Ralph, had a bicycling accident that left him a C-4 quadriplegic, paralyzed below the shoulders.

He was 55 years old, a graduate of Caltech, and a nuclear physicist. I was a 42-year-old childless jock who worked for an adventure-travel firm leading bicycling trips to exotic locations like Tasmania and Bali.

The only things I ever wrote were postcards, grocery lists, and, occasionally, overblown copy for the company's travel brochures. But, in the spring of 1994, after Ralph's accident, all forms of writing became obsolete. I spent my days dealing with doctors, therapists, and social workers. At night, I lay in bed alone, wondering what would happen to us. Ralph and I were now companions only in the sense that he needed me for eating, voiding, and changing his position, the sheets, and the channels on the TV. I thought I would go crazy.

While Ralph was still in ICU fighting for his life, a friend advised me to write down everything that was happening to us. He thought that I might need these notes for a lawsuit. He bought me three spiral notebooks and a pen, and I dutifully jotted down what I saw, heard, smelled, and thought. This may have been the beginning of my writing career, but once Ralph returned home, I never referred back to the notebooks. I knew they were full of bleak thoughts about running away, murder, and suicide. Years later, when I found the notebooks in the corner of a messy closet, I threw them away without looking at them. By then, I had written a memoir about our new life.

Six months after we returned home from the hospital, I began to keep a journal: I scribbled down the things that happened to us—what I saw with my new eyes as a stay-at-home caregiver; who I met; how the world reacted to our new status as a disabled man and his helper. I was encountering people who I never would have run into before: neurologists and psychiatrists, acupuncturists and kinesiologists, African American and Jamaican neighbors, and out-of-luck health-care "professionals"—the people I had come to depend on for my husband's care and for my own sanity.

We were in and out of the emergency room so often that I started to think of it as my office. Waiting there took an average of seven hours. But I always had a notebook and pen with me. ER became my home-away-from-home.

At our house, we endured visits from social workers, home-health nurses, therapists, and aides; each of them had an agenda and an idiosyncratic timetable. Then, as abruptly as the accident itself, our HMO stopped providing us with home support; we went from multiple helpers to only me. But taking care of Ralph was not a job I could do alone. We needed 24-hour, live-in assistance. However, our insurance did not cover this necessity, so we were forced to become entrepreneurs.

Ralph and I went into the business of keeping him alive and making ends meet. I became our house manager, hiring and firing attendants at a dizzying rate, deciding which flaws we could ignore and which we couldn't. We went through a succession of drug addicts, alcoholics, thieves, kleptomaniacs, and pedophiles, until finally we found a semi-stable, live-in helper named Jerry, whose only visible drawbacks appeared to be an addiction to gambling and a very bad driving record. Much later, I discovered that he had spent five years in San Quentin for statutory rape, pimping and pandering.

As a recently retired scientist, with 25 years at Lawrence Livermore National Laboratory, Ralph was more comfortable with numbers than with people. Now he was often so overly medicated that he was confused by the constant flow of humanity in and out of our home. I was in charge of everything: his care, our finances, our non-existent social life.

I became best friends with a 300-pound, polyester-and-sequin-swathed neighbor, Mrs. Scott, who took over our kitchen. She kept us supplied with fried chicken, greens, ham hocks, and pound cake. She was my confidant, spiritual advisor, therapist, and surrogate mother. There were moments when I devoutly wished to curl up in her enormous, soft lap and go to sleep for a very long time. There were other times when I wanted to do to her just what she threatened to do to me: slap her upside her big nappy head and knock her into next week or Christmas, whichever came first.

Obviously, I had things to write about. And, oddly enough, I found the time to write: late at night and early in the morning, when Ralph was asleep; on our trips to ER and in the waiting rooms of doctors and counselors. Since we didn't have a social life and I no longer ran, biked, skied, or climbed, I substituted writing for friendship, exercise, and sex.

It took a while for me to let go of my identity as an adventurous jock. To keep some money coming in, I took a full-time job—with flexible hours—at an indoor climbing gym. I enjoyed being surrounded by nubile, athletic bodies, by people whose only worry was whether or not it would rain on their next climb up El Cap. I liked—and needed—this contrast with my new existence as Ralph's care-giver and companion.

I surrounded myself with chaos, because the more diverse, oddball people I let into my life, the easier I found it to cope. Mrs. Scott and Jerry provided me with help for Ralph, but they also gave me a multitude of distractions. Now I not only had my husband to worry about, I had their needs to think of, too.

I had grown up near Philadelphia on a mini-farm. My father was a biologist; my mother was a homemaker. My parents were college-educated, Presbyterian-Republican members of the country club. (Actually, my father raised rats for a living, rats to be used in research, thousands of them. I was the most popular kid in grammar school, because every year my class went on a field trip to my house—the business was located across the driveway. Also, my father gave jobs to every teenage boy in town; they'd last about three months, because no one really likes to clean rat boxes: Rats stink and sometimes they can be mean.)

I was a good student and a dutiful daughter. I spent summers with my three brothers and my mother's family at the Jersey shore. In grammar school I took piano, ballet, and golf lessons. I went to an all-white high school and, in 1970, I entered an expensive, conservative, private college near Valley Forge, PA. It was my first taste of living without the constant shelter of my generous parents. I went through a small, rather unillustrious, post-hippie rebellion. Instead of majoring in English, I studied sex, drugs, and rock 'n' roll. I dropped out of Ursinus, hitchhiked up the California coast, went back to New Jersey, and enrolled in a state school near my parent's home. I graduated in 1974, with a degree in elementary education. I took a teaching position in Lexington, VA, the home of Virginia Military Institute and Washington and Lee University and the final resting place of Robert E. Lee, Stonewall Jackson, and both of their horses. I married a hometown boy, a graduate of Washington and Lee Law School. When that didn't work out, I relocated to San Francisco, where I should have moved ten years earlier.

I met Ralph my first year in California, on a bicycling trip in Baja. It had been advertised as an "alternative" adventure. It cost $175 and consisted of 17 days of grueling bicycling, crisscrossing the peninsula, from Tacate to Cabo San Lucas. There were 45 participants, four to a room, two to a bed. Because I was the last of five single women to sign up for the "vacation," I was assigned a male bedmate. Many people dropped out along the way, including the stranger I was sleeping with. (I assumed it was the days, not the nights, that were too intense for him.) To economize, the organizer consolidated rooms and roommates. One night, Ralph was rotated into my bed; he never left.

From that day onward, we were a couple, and, eventually, after ten years together, we got married. I was 40, Ralph was 53. I married him because he could read a backcountry ski map and set up a tent in the snow. He married me because I could keep up with him. It was not an intellectual relationship, and we didn't talk about our feelings and emotions. We DID things together like planning bike trips and dinner parties.

We searched for a year until we found a house that suited both of our tastes. Ralph loved Victorian architecture and everything Art Deco. Our 1907 Queen Anne, in a predominantly African American neighborhood in the flatlands of Oakland, was in fairly good condition, with many of its original features, including hardwood floors, fanciful wallpaper, and detailed woodwork. Ralph approached house restoration as he did everything else: with compulsive gusto and almost excessive enthusiasm. He put all his energy into renovating windows and refinishing floors. He took woodworking and stained-glass-making classes. He made our home comfortable and beautiful.

We were too busy fixing up the house to get to know our

neighbors. But after Ralph's accident, we never went anywhere. We were always at home. And when we attached a wheelchair ramp to the side of our house, some of our neighbors, including Mrs. Scott and Jerry, strolled up to introduce themselves.

I had grown up not knowing any black people, except my mother's cleaning ladies. Now I spent more time with Mrs. Scott and Jerry than with anyone else. I learned that Mrs. Scott had been raised on a sharecropper's farm in East Texas. I found out that Jerry had been bounced between various relatives' homes up and down California's Central Valley, until, finally, he wound up on the streets of San Francisco, stealing cars and running with women of questionable repute.

I realized that I was in a unique position. These were the people I was living with; I depended on them for my very freedom; I was not a social worker or a probation officer who saw them occasionally, by appointment only. We shared a bathroom and a kitchen, and, eventually, I shared Jerry's bed.

Sleeping with Jerry was troublesome for me, but I needed comfort, and he was available, willing, and (very) experienced. Crawling into his bed was convenient. He knew better than anyone else the situation Ralph and I were in. He understood me, more than I understood myself. His entire life had been spent working for white people—as a mover, a mechanic, a janitor, and home-health worker—whereas my life had been lived in the company of people just like myself.

I listened closely to Jerry and Mrs. Scott—I had to—and I jotted down their thoughts and sayings. Mrs. Scott was sure that God could cure anything. Jerry knew that only money, lots of it, fixed what ailed you.

I did not show my writings to anyone for a long time. About a year after the accident, Leah Garchik, a reporter and columnist at the *San Francisco Chronicle,* asked her readers for stories about "serendipity." I sent her a description of a serendipitous experience Ralph and I had had. The ambulance driver who had driven Ralph to the hospital after his accident had tracked us down. He came to our door bearing carnations. He told us that Ralph had been very brave as he lay in the middle of Claremont Avenue in Berkeley, unable to move. He explained that he had not thought Ralph would survive. It was a poignant meeting, and my piece reflected the intensity of our emotions. Leah wrote back to me and said that my essay was too long, but that it was very good and that I should send it to her editor. This is how my publishing career started.

I didn't send the piece to Leah's editor, Andrea Behr, because I had no clue as to how to do so. But I told Leah that I had more pieces—about accompanying Mrs. Scott to the grocery store, Ralph to ER, and Jerry to the police station. When I mentioned that I had stories that were graphic about sex, she asked to see them, clarifying that she was not a sex maniac, but that it sounded interesting. She read my scribblings and encouraged me to write more. She thought I had the makings of a book.

She invited me to visit her office and showed me around the newsroom. She took me to the M&M Bar, where the old-school reporters drank. She marked up my writing and gave me advice. She introduced me to her husband and invited me to their Christmas party. She called me at least once a week to check in, to listen, to empathize.

Leah introduced me to Bonnie Nadell, David Foster Wallace's agent. Bonnie took me on as a client. She was attracted to the sensational qualities of my story, that is, sex with Jerry. I had not yet told Ralph I was sleeping with Jerry. Bonnie suggested that in order to complete the book I needed to confess to Ralph. Suddenly, life was beginning to imitate art. I wasn't ready to make that leap. Bonnie kept me on as a client, but not for long.

We called my manuscript "Tumbling After," from the Jack and Jill nursery rhyme, and Bonnie sent it to six of the largest publishing houses in New York. It was promptly rejected. Bonnie unceremoniously dropped me and advised that I keep living my life. "In about ten years," she said, "you may have a story." I had expected her to stay on as my agent, and I was devastated. But Leah never faltered as my mentor and guardian angel. She prodded me onward, giving me counsel and confidence. Without her help I would have given up.

I enrolled in a basic writing class taught by Berkeley journalist and activist Meredith Maran. Then I took another class from Adair Lara, a columnist at the *Chronicle*. And I signed up for a UC-Extension course taught by an ex-*Chronicle* columnist, Gerald Nachman. It was called "Writing with Humor," and it was an excruciatingly unfunny experience. During coffee breaks, I would run down to the corner bar and buy myself a glass of wine—my only way of coping with Gerry's teaching style. But, as in every class I attended both before and after his, I did each assignment and attempted every exercise. I wound up with a large body of work. The only trouble was I didn't know what to do with it.

I bought a subscription to *Poets & Writers* and circled everything: contests to enter; conferences to attend; anthologies, magazines, and websites to submit to. I applied for workshops that offered scholarship money. In the summer of 1998, I went to three conferences: Aspen Writer's Workshop, Port Townsend's Writers Week, and The Hurston/Wright Writers of Color Week. Each place provided me with a generous discount.

Ralph was supportive of my ambitions. He thought we might grow rich from my writing. Meanwhile, attending these conferences made me more self-assured and determined. I relished being around other writers, and away from Mrs. Scott, Jerry, and Ralph. Back at home, I began to look for new friends, folks who were serious about literature and good writing, people who appreciated my voice and sense of humor.

The Hurston/Wright conference taught me that my writing could incite strong emotions from readers. The other participants didn't like my manuscript; they asked that I be removed from the memoir class. They said I stereotyped blacks and used reckless, inappropriate language.

In the fall of 1998, four years after Ralph's accident, I finally had an essay published in the *Chronicle*. I continued to take classes, send out essays to magazines and newspapers, and work on my memoir. I was published a few more times, in the *East Bay Express,* the *Washington Post,* the *Chicago Tribune,* and *The Sun* and *Hope* magazines. Whenever I was rejected, I immediately sent another piece to the same editor. Sometimes this strategy worked, sometimes it didn't. After nine submissions and subsequent rejections, *salon.com* published one of my essays about envying the bra of a nine-year-old neighbor. (It had a picture of Daffy Duck on it. Who wouldn't envy that?)

An editor from *salon.com* sent me an e-mail requesting that I re-send her an essay about partnerless sex. It was obvious that she had confused me with someone else. I hadn't sent her anything, and I'd never written about sex without a partner. But that didn't stop me. I quickly wrote something inappropriate and submitted it to her. I never heard from her again.

In 1999, I won the Richard J. Margolis Literary Prize—a stipend of $2000 and a month-long stay at Blue Mountain Center, an artist's retreat in the Adirondacks. The prize was for giving voice to populations who are not always represented in literature. It took me two weeks to get used to the absolute quiet of those gorgeous mountains. I loved having no responsibilities except to write.

When I returned home, I applied for a residency at the Headlands Center for the Arts. I was awarded a Bay Area Artist's Fellowship, which included room and board at a converted army barracks located along the rugged coast just north of the Golden Gate Bridge. It was an extraordinary experience. I was surrounded by painters, sculptors, dancers, musicians, activists, and philosophers. It helped me see myself, for the first time, as an artist.

I acquired a new agent, Amy Rennert, of Tiburon. I found her name in the Learning Annex catalog, but I remembered her from long ago, when, as the editor of KQED's *San Francisco Magazine,* she had taken a Canadian Rockies bicycling trip with the company I worked for. With Amy's help, I shaped and reshaped my story, edited and rewrote, and edited some more.

Of the 13 editors Amy and I initially sent the manuscript to, there was one that I had met beforehand, Carol Hauck Smith, Pam Houston's and Ron Carlson's editor at Norton. I knew she would remember me, because she had read the beginnings of my memoir at the Aspen Workshop. As it turned out, my manuscript came close to being accepted at Norton because of Hauck Smith. It was a lesson in the value of self-promotion. To make it as a published writer, you must be good at the craft and have confidence in yourself, but you must also be willing to mix, mingle, and market. Even so, I was rejected at Norton and everywhere else.

Amy and I reshaped the book again, rewrote the cover letters, and came up with a game plan. I had a difficult time writing a proposal and at the last minute, at Amy's suggstion, I hired Marilee Strong, author of *A Bright Red Scream: Self-Mutilation and The Language of Pain,* to help me. Although I thought of myself as a self-promoter, I just could not get the lingo down that would draw the attention of New York editors. Marilee had sold her book based on a magazine article. She was good at researching facts and figures, something I had no aptitude for.

We submitted the proposal and 50 pages of the memoir to six new publishers. Just when I began to wonder if there were anyone else left to submit to, Crown offered me a contract. It was not a $100,000 deal. It was not even a $30,000 offer. But Amy recommended that I accept a $25,000 advance, and I did. My dreams of becoming rich were rapidly fading, but I gave myself a party in the backyard to celebrate; Jerry barbecued.

The acquiring editor, Betsy Rapoport, had been drawn to my manuscript because her brother-in-law had had a bicycling accident several years before. Although his injuries were not as extensive as Ralph's, Betsy knew first-hand how his accident had affected her sister. Again, I was struck by how much chance plays a role in getting published. I was lucky to have found an editor with a personal connection to my story.

Betsy asked me to make many changes. She deleted chapters, moved characters around, and insisted I update it with the latest events in our house. These included Ralph's drug overdose, caused by two doctors changing his meds without consulting one another; a stint in jail by Jerry; a new boyfriend for Mrs. Scott; and the addition of another caregiver/housemate, Harka, who came to us, at the recommendation of friends, from Nepal.

Harka's innocence and enthusiasm for everything American were both charming and frustrating. He butted heads with Mrs. Scott. He was afraid of Jerry. He fell irrationally in love with me. But he added a fresh perspective and a different dimension to our lives. He gave me a delightful new character to write about.

The most difficult part of the editing process was adding personal details about Ralph and me. Our relationship, or lack of same, was what the previous publishers had found disconcerting and what eventually led to their rejections. Betsy and Amy forced me to face this challenge. Like no therapist before them had, they made me look objectively at my marriage. I had never been a reflective, analytic person. Ralph and I did not have an intellectual relationship. And, even after his accident, when his body was useless and his dependence on me overwhelming, we didn't dwell on the past—or on our feelings. We concentrated on the day-to-day monotony of getting him bathed, fed, and up and into his wheelchair.

By pushing myself to write about emotions I did not want to confront, I realized that I still loved Ralph. I didn't love him in the way I had once loved him, for his independence and athletic prowess. Now I loved him for his mental toughness and incredible emotional strength. I understood that I was lucky to be married to a man who did not complain about his loss and misfortune, but only focused on the here-and-now, what he could do, not what he couldn't. And I knew I loved Jerry and Harka and Mrs. Scott, too: Jerry for his sense of humor and laissez-faire attitude, Harka for his faithfulness and innocence, Mrs. Scott for her help and misconstrued guidance.

Slowly, Ralph became more independent. He got involved with the Berkeley Center for Independent Living, becoming a board member and working on advocacy issues. Not long ago, he found himself stuck on an elevator at the Oakland Coliseum, in part, a planned one-man demonstration to draw attention to inaccessibility problems.

With the aid of the Internet, Ralph dabbled in the stock market and made some money. He devoted his free moments to trading, and he was in the right place at the right time. Eventually, he made enough so that I could temporarily quit my full-time job. I worked on my manuscript for another year. I won a few more contests. I began teaching writing workshops.

I got published semi-regularly in the *Chronicle*. I was happy when I was in the paper and depressed when I wasn't. The more I was in print, the more I learned that my writing caused controversy and debate. From the Letters to the Editor after each of my columns I learned that the disabled community thought my depictions of Ralph were insensitive and inaccurate. There were people who thought I was bigoted and racist, whiney and clueless. I received thousands of e-mails, most of them complimentary, but it was the ones that were full of criticism and outrage that stayed with me. On the days when I received no negative comments, I was disappointed. It turns out that I like making readers angry and uncomfortable!

Then the *Chronicle* stopped buying personal essays from freelancers. I had my car towed to a junkyard after Jerry totaled it. Ralph is not doing well in the market, and I'm down to cashing in IRAs and paying only the minimum on our credit-card bills.

But still, there is no doubt—writing has saved my life. □

Susan Parker lives in Oakland and writes a weekly column for the Berkeley Daily Planet. *Her memoir,* Tumbling After, *was published by Crown. E-mail: sqparker@pacbell.net*

MIDNIGHT IN THE MAZE AT GRACE CATHEDRAL

John Ryan

I spend half my life near the intersection of Irving Street and 9th Avenue. I get my morning bagel at Noah's, my large coffee at Starbucks. I rent laserdiscs at Le Video near Lincoln, eat chicken teriyaki at Ebisu, fish the *New York Times* out of the box by the Rexall. And I try not to get attached to any of the people I come in contact with, because this place is really transient. I suppose salaries aren't high enough to keep anyone with any sense of self. So even though I order the same Grande Drip and Odwalla orange juice every morning from that tall blonde at Starbucks, and even though I get the same poppy-seed bagel with light plain cream cheese every morning from the black gay guy at Noah's, and even though I've asked that freak at Le Video, the guy with the Lennon specs and Zeppelin tattoos, to hold *Battleship Potemkin* or *Jerry Maguire* for me at the counter, I never pay attention to their name tags or engage them in small talk. Because they're all going to be gone soon.

My UPS man's name is Clyde Wheeler, and he has a daughter that just got a free ride to Princeton for math. I know this because we talked last December every time he dropped off a Christmas present—an Odyssey putter from my dad, a barrel of popcorn from Aunt Gladys, some Chinese tea-can lamps from my rich brother in Atlanta. The UPS man calls me Henry and I call him Clyde and he knows the whole story about me getting fired and about my boss. This is a guy I saw maybe five times in the last three months.

But on Irving St., forget about it. They won't be at those places very long, so I repeat the same monotonous orders to them every single morning.

I spend a fair amount of time walking around town, because,

even though I really love my apartment in the Haight, I sometimes have a hard time thinking clearly when I'm there. Often I just walk straight out my door, up Clayton, up the hill to Twin Peaks. They have those big steel 25¢ public binoculars up on top of Mt. Sutro, so I always make sure I've got a lot of quarters in my pocket. If you look out to the west, you can watch the rain come in off the Pacific. You can't see the tiny sandpipers that race around the pools of the outgoing tide, but they are there. On a really clear day you can see the Farallon Islands and the occasional whale. To the south is the Stick and the airport, and Glen Eagles where Ed and I play golf. You have to get off the fourth tee pretty quickly at Glen Eagles, because to the left is a chain-link fence. And on the other side of that fence is the Sunnydale housing project. People have gotten their clubs, wallets, and watches stolen on the fourth tee at Glen Eagles.

To the east, down Market St., is the perfect urban bombing run. Right down the pipe behind the wheel of an F-16, all I'd need to neutralize all those banks and Gaps and Macy's annexes, and the Planet Hollywood, and the Hearst Building, and the San Francisco Port Authority, and, of course, the Bechtel facility on Beale St. where I used to work—all I'd need would be half a dozen Sidewinders. Really, they should mount a howitzer up at Sutro, so you could slide in a couple of coins and lob some artillery downtown. But I suppose that even if I was lucky enough to wipe out all the bike messengers and the securities traders and the Unix administrators and the copywriters and the legal secretaries—even if I could send a few .50-caliber shells through the front doors of Hermès or Au Bon Pain or Saatchi & Saatchi or *Wired*—I'd still feel a little guilty about burying Andrea, my former boss, in a pile of rubble.

She called me into her office one day last June—this was back when things were still good between us—and she proceeded to go through my yearly performance review. I was a unit manager for the Blue Group, our Chilean mining division, and I had really kept things running smooth. I thought that Andrea was just going to jerk my chain throughout the meeting—we were still in that puppy-love stage of our relationship. But, surprisingly, she really had a lot of criticism of my management style. Nothing serious, but she told me that I needed to work on managing laterally.

She was accurate in one respect. Blue Group was a hard-dollar operation, which is to say that the tungsten we pulled out of the ground made Bechtel money. Most of the other San Francisco divisions, like Research or Systems or Implementation, were soft dollar. Sure, they were equally important in the grand scheme of things, but they weren't profit centers like Blue Group. So, as you might imagine, my department got a lot of perks that the other guys didn't. We got to go to Maui for our corporate retreat, for example, and I had a pretty generous expense account, a membership at the Olympic Club, and, considering it was Bechtel, a pretty sweet office that overlooked the Bay from the 15th floor.

When Andrea told me to manage laterally, I saw what she was getting at. She was telling me to chum it up with the other managers. And take the snapshots of the group windsurfing in Maui off the breakroom bulletin board. I got the concept, but I was still stung. I didn't like hearing criticism about my personality when Blue Group's numbers were looking so good.

Andrea must have caught that defensive look in my eye, because all of a sudden she got quiet. For a moment, I forgot she was my boss; I expected her to say something sweet and encouraging, like "Don't sweat it, honey. Just tone it down for six months." Instead she said, "Look, Henry, we're not running a daisy farm here, but you've got to work on your attitude with Bill and Christine and Sun Ji. You can't keep taking the engineers to Boulevard and Aqua at $500 a pop. I don't care how bright the figures look, it's not the way we do things here."

I could hear the blood pumping through my ears and I could picture her little body hurtling toward the plaza down below. There was a lot I wanted to say to her, but I'd been in the business long enough to know better than to speak up when I was pissed off. I said thanks and got up to leave.

"I love you," she said.

It was the first time she'd said it. I could tell she really meant it, and I was not moved. I knew I didn't feel anything for her and never would again. I don't know why this happened, but I think it happens to a lot of people. It was an epiphany.

Epiphanies happen at strange times, although they often happen in familiar places. Last week I was walking down Fillmore a little after midnight, high on coffee, just putting one shoe in front of the other.

As I walked along, I was overcome with a profound bitterness. All the little shops were tricked out for Christmas—lights strung across window frames, fake snow sprayed in the corners, holly and poinsettia hung in the low eaves. There was so much overpriced crap for sale that I started to hate everyone, especially everyone with a job and a family.

All the quaint shops were packed with green candlesticks and tortoise-shell vases, Bakelite soap dishes and woven baskets of flavorless bon-bons, specialty coffee beans, pink tea towels, painted flower pots, gingerbread houses, handmade Native American nativity scenes (as if). . . .

And I couldn't find a place to sit down. All the benches and café tables had been packed up and wheeled in. So I walked, and the bitterness began to work its way into me pretty good.

That's when I ran into Mary Marlowe and Chrissy Something or Other. They were coming out of Jackson Fillmore—a retro-yuppie trattoria at the corner of Jackson and Fillmore, n'est-ce pas. I went to prep school with these girls back in Connecticut, but I didn't really remember either of them. They looked like they were doing well, dressed in black, their hair done like they were at a party. They were on me pretty quick, asking what I was doing in Pacific Heights at midnight looking like some wild-eyed freak. It all happened so quickly that I couldn't think of any good reason why I was there. So I went with the truth. I told them I was taking a walk. At midnight? I'm an insomniac. Drinking coffee? Why fight it.

They reminded me that they had roomed together in South House my senior year, and they seemed pleased that they still lived together ten years later. I doubt they were interesting enough to be lesbians, which made me feel sorry for them. They had to be pushing 30, and neither one of them had found a husband. That sounds sexist, but I'll bet their mothers are getting worried about them as well.

They told me they were both in advertising. Mary was producing radio spots for Safeway at McCann-Erickson. Chrissy was an account exec at Goodby, Silverstein. Mary was leaving in the morning to go on a team-building retreat at some shithole in Monterey. She said her boyfriend was going to try and join her, but more than likely Chrissy would come down Saturday afternoon to enjoy some spa treatments on McCann's tab.

Then they started in on me. They wanted to know if I kept

up with Matt Freuhauf, with whom, according to them, I had been best friends. By their account, he was hawking bonds at Salomon Brothers and had married a girl named Claudia, the daughter of a federal judge.

What was I doing? I told them I was out of work. Retired, I joked. So then Mary told me she'd always had a crush on me, but that she'd been too nervous to pursue it. She'd been intimidated by me, whatever that meant. Telling me about her crush seemed to give her a big thrill. I suppose I wasn't very intimidating anymore.

Before they could sneak in any more jabs, I told them I had lung cancer. That shut them up. I lit a cigarette and they went into a deluge of consolation.

They said their good-byes quickly, adding that they hoped to see me up in Tahoe. I wondered how they could say such a stupid thing. Tahoe? For a moment I wished I'd known about Mary's silly crush when it could have done me some good, like senior year.

While I headed down Jackson past Danielle Steele's mansion, I felt guilty because I'd lied about having cancer. I crossed over at Taylor, and eventually found myself on the steps of Grace Cathedral. I thought I might pray for my soul.

Even at midnight, there were a few stragglers walking the maze that is darkly tattooed into the light granite pavement of the terrace near the rectory. It's about 40 feet in diameter, a series of loops that snake inside each other around a cloverleaf. Of course, you can walk straight to the center, but I've seen a lot of people walk this intestinal tract for five minutes or longer, careful never to step on any of the lines, often stopping to look back. They seem to need to make sure that they haven't lost their way, unlike lab rats who whip through mazes to earn food pellets or avoid electric shocks.

An older couple was leaving as I arrived. "Excuse me," the guy asked. "Did you get that coffee around here?" He was wearing a Cincinnati Bengals fanny pack and his glasses were slipping off his nose.

"No," I said. "Over on Fillmore about ten blocks that way."

"Strange to hear a Southern accent," the woman said. "Do you live here in San Francisco?"

"Yeah, but I grew up in Georgia."

"Not Savannah?" she asked, smiling.

"That's right."

"Well, have you read *Midnight in the Garden of Good and Evil?* If you haven't, you'll just love it."

"I have, of course," I said, trying to be polite, although I don't believe you have to be polite to tourists in the small hours of the morning. They should have been resting up for an early walk in Muir Woods. I could have mugged them and been well within my rights.

"You must tell me," she went on without restraint. "Was it really as weird as the book says?"

I kind of laughed. I figured I was into it by now and couldn't in all fairness switch gears, so I launched my standard monologue, about how there were so many tour busses all around the last time I visited, et cetera. Et cetera.

"Well, I'm just dying to go, but this one," she said, cocking her thumb toward her husband, "he'd be happy if he never left Cincy again."

"Well, it was nice meeting y'all," I said. Then I walked up the stairs, back up to the maze.

"Did you hear him say 'y'all,' Henry," she said. "I just think that's so precious."

They took the stairs down to Taylor very slowly, and I watched them head past the Pacific-Union Club. I went over to the entrance of the maze. I looked around to make sure no one was watching me. And I walked into it.

These days, I can't tell anybody that I grew up in Savannah without having to field a barrage of questions about *Midnight in the* et cetera, et cetera.

I was 13 when Jim Williams shot his hustler boyfriend, Danny Hansford, at Mercer House, so I don't really remember that much. First of all, I didn't even know that Mercer House was called Mercer House, just as I'm sure no one in Palm Beach called the Kennedys' house the "Kennedy compound" before it ended up on CNN. It was just Jim Williams's house and it sat on the west side of Monterey Square. We used to play in Monterey Square, because it was one of the few squares that had been closed to traffic.

Back then, traffic used to cut right through the squares, which was dangerous for kids playing, but more effective for fire trucks. Savannah was, and still is, a tinder box of woodframe Colonials and Victorians packed full of four-poster beds and sprawling drapery and heart-pine floors and ancient gas heaters. Watching those massive trucks trying to get around the squares on their way to a four-alarm now that the squares have been closed off is a joke. But the squares look good and tourists come to walk through them.

In the early eighties, the historic district in Savannah was still a burnt-out Dodge City, a hodgepodge outpost of eccentric young rich people and assorted other bohemians who mostly partied their asses off. Not many houses had been restored, and those that had been were occupied by people like us, who kept their doors locked tight and put security tape over the windows. I remember being jealous of friends who lived on the islands, or near the yacht club, or in Ardsley Park to the south, or in Habersham Woods out near the movie theaters—the good neighborhoods. Their parents wouldn't even let them come spend the night at my place, because they knew about all the creepy shit that happened—people got shot, bicycles got stolen, bars let twelve-year-olds get ripped on Russian Quaalude shooters. Oh, sure, and people had nice houses.

We were just kids and we used to hang out at Jenny Long's house, or Smithy McIntosh's, because he had a pool. And, outside, in the street, people were screaming around the squares in Camaros, and 82nd Airborne Rangers from the local army airfield were beating homosexuals to death in the quaint little parks. And, yes, there were a lot of azaleas.

And there were a lot of rivers that looped around outside of town in the intercoastal waterway. And they used to have a "long ball" contest down at Spanky's on River St. to see who could drive a Titleist over the river and onto the docks at Hutchinson Island.

And there were a lot of guns, and one of them belonged to my stepfather.

Anyway, I remember when they were filming the movie *The Lincoln Conspiracy* and Jim Williams hung the Nazi flag off his balcony. No matter how many times it was explained to me, I couldn't grasp why he'd go to such extremes to keep his house out of the camera shot.

As far as I knew, he was a Nazi.

I remember my parents coming home from one of his Christmas parties one year, which is odd in itself, because they went to a lot of parties and it's strange that I remember this particular one. Williams was living with an interior decorator named Philip, and my mother was especially impressed with the delicate sprays of orchids that were hanging from the valances and curtains throughout the house. She said she'd never seen anything like it. As far as I knew, we had really nice curtains at our house.

My stepfather didn't say much about the party, as I recall, and I imagine that was because he considered Jim Williams to be quite the shifty scumbag. No one really believed that Williams made enough money from his real estate transactions to buy all the Fabergé boxes and other French crap that he had scattered all over his house and in the basement. There was talk, but nothing concrete.

My other brush with *Midnight* happened about five years after Williams shot Danny Hansford. I was 17 and had been kicked out of Canterbury in the spring of my junior year for reasons that were, more or less, beyond my control. The summer after sophomore year, a letter was sent announcing a schoolwide ban on tobacco products. This meant no smoking and no chewing tobacco. Friends of mine in Savannah thought it was hysterical that some preppy ghetto like Canterbury would have to outlaw chewing tobacco. None of us ever chewed, we just smoked it like normal people. I remember a group of us sitting around my house the day that letter came, and, after Bill Hyde read it, he said "What kind of rednecks are they letting into Connecticut High School these days?" They all referred to Canterbury as Connecticut High School.

"All the jocks dip Copenhagen," I said. I was staring at the letter and probably trying to think of all the places on campus I'd be able to sneak a smoke during the next year's prohibition.

"Good God," Bill said, laughing. "I feel sorry for you having to board with those Greenwich hillbillies."

When I went back to school in the fall, it wasn't long before I was caught lighting up in the laundry room of my dorm, and in the A/V closet in the library, and at the pizzeria down by the barber shop, and in the barber shop itself. By spring, I'd accumulated so many probation hours that I was campused. That meant no weekends in New York, not even any walks into town, where I had a much better chance of not getting caught smoking.

By the time Mrs. Mandler caught me huffing a Marlboro in the boiler room, I was ready to head home. You might think prep school kids have it made, but the truth is: Being confined to campus is torture. Friends of mine back home might have lost their phone privileges or gotten grounded for their fuck-ups, but at the end of the day, they got to go home. I was under house arrest.

So I was thrilled about going home, where I'd be able to smoke all the cigarettes I wanted. I was also quite excited about the possibility of drinking vodka all night on Saturdays, puking in my friends' cars, and screwing cheerleaders! Canterbury didn't have cheerleaders, because the girls had their own teams to play for.

The school agreed to let me finish out the year's work at home, taking an occasional test they sent to a tutor my mom had set me up with. I had to go to her house three times a week for three hours a day. More than a few of those times, I'd go in reeking of booze. She was compassionate, which was a break. After we'd talked a while about Willa Cather or the Mayflower Compact or the water cycle, she'd let me take a nap on her couch.

By the way, JFK had gone to my prep school and gotten kicked out! And Dominick Dunne went there, too, and he's rather famous for writing about rich people that kill and get killed.

So I took to raging pretty hard that spring, staying out until dawn with my older brother, Angus, and his friends—drinking heavy amounts of Budweiser and lemon-drop shooters, and gobbling up various types and quantities of drugs. Under the circumstances, I suppose it was inevitable that I would end up at Ernesto McCoy's house one Thursday night at four a.m.

Ernesto McCoy was a personal injury lawyer who had long since disabled any braking mechanisms he'd ever had on his life. I met him in the Lamppost Saloon on Bay St. one night when I was shooting pool and watching the fat Korean strippers. Ernesto came in with Joe Odom and they were wasted, but functional. Bill Hyde and I played them and I ended up shooting the eight ball on the corner pocket.

Just as I was getting ready to stroke it, Joe gave one of the strippers five bucks to rest her tits over the pocket. I put the ball in slow, and Bill and I drank another round on Ernesto and Joe.

When the Lamppost closed for the night, we all walked back to Ernesto's house on Oglethorpe St. He lived in a gorgeous three-story brick town house on the corner of Whitaker, and his front door was twice as tall as I was.

In the living room, Ernesto showed me a picture of himself and some huge black guy sitting in deck chairs on some tropical beach. In the picture, Ernesto had a plastic wastebasket on his head. "That's when I was crowned King of the Bahamas," Ernesto said. "That big black bastard right there is Julian Tryst, the biggest dealer on the island."

Ernesto put the picture back on the mantel and picked up a small Chinese bowl. It was half filled with pink cocaine. Ernesto worked his way around the room, letting everyone get a nice noseful.

In the course of the conversation, Joe Odom mentioned he'd been playing a lot of chess lately, so I offered to play him. We played four games. I destroyed him. Then Joe got a serious nosebleed and had to take off for the bathroom. Ernesto stepped in and I destroyed him, too.

Around six a.m., Bill was cracking some ice for a fresh drink, when he slipped and stuck an ice pick through his hand. So we left.

Anyway, my point is that what Berendt missed when he was doing his research was the dynamic of the scene. His readers might think Savannah was this hotbed of sin and high society and thick bloodlines, but, at least the way I saw Savannah back then, it was infinitely more graceful and elegant than Berendt was willing—or able—to admit. He was an outsider for starters, and I think he became too easily mesmerized. No one downtown knew who Lady Chablis was, or Miss Minerva, or that guy with the flies attached to his shirt buttons who was going to poison the water supply. I'm sure Berendt could have found a serial killer in Savannah, or group of radical fundamentalist Christians, or a character like Forrest Gump—you can find these folks in any town if you look in the right places. It may seem like heresy, but I think that Lady Chablis and the voodoo priestess should have been cut. I think the story would have been more accurate without them.

The only other relevant connection I have with *Midnight* is that a year and five months after Jim Williams shot Danny Hansford, someone broke into my house and shot and killed my stepfather while I was sleeping upstairs.

There was blood everywhere. Not just on the TV-room floor, where it spread out around his body like an old sleeping bag. It was in his salt-and-pepper hair and under his fingernails and on the dust ruffle of one of the chairs, and on his teeth, and on the cover of *Esquire,* right up to Norman Mailer's chin. It was on his .357 Magnum, which had found its way across the room over by the French doors, and it was on the gloved hands of the county coroner, who was filling out some paperwork using the top of the

TV as a makeshift desk. It was also on his pen, and it was on the TV, too, it had pooled up around one of the legs of the old black-and-white Sylvania.

A few spots had spattered on the 14-foot ceiling, and it was on the soles of the Savannah policewoman's heavy black shoes as she walked me out of the room and sat me down on the red loveseat against the far wall of the dining room. Two Georgia troopers were trying to wash the blood from their hands in the little bar-sink by the kitchen, so there was blood in the sink and on the Ivory Liquid bottle and on the hot and cold handles.

It's only important to psychiatrists and homicide detectives that I saw all this blood on Halloween morning, exactly one month before my 16th birthday. Therapists always want to know how I felt when I saw the hole in my stepfather's chest—the breast pocket of his blue plaid bathrobe scorched and blown out by the gun blast. The detectives just wanted to know if I'd heard or seen anything suspicious that morning.

I'll just tell you what I told them: I didn't feel anything. I hadn't heard anything. I didn't see anything, except all that blood.

A few days after the murder, we moved into Charlotte Darlington's house on Oglethorpe St. across from the Colonial Cemetery. It was an enormous three-story eggshell Georgian with a great parlor, huge cherry sleigh beds, and cavernous armoires in the bedrooms. Mrs. Darlington was in her nineties. Because the big house had been too much for her, she had moved into the carriage house about ten years earlier. She had had a collection of housekeepers that kept the main house tidy and aired out over the years, and there were always fresh flowers on the hall tables and in the window boxes that hung outside over Oglethorpe St..

Before she relocated to the carriage house, Mrs. Darlington had installed mechanical chairs that ran up along the sides of the staircases. I used to ride them up and down whenever I was alone. They were very loud, so I rarely took a spin when my mother was around.

About a week after we moved in, my mother came home from the grocery store and told me that there was a slight chance I might have to take a polygraph. The police had officially ruled out suicide, which, for obvious reasons, was a huge relief. However, since the case was now considered a homicide, I was a suspect. My prints were on the gun and the police had found some drops of blood on the bed skirt of my bed. Many things from my room were being analyzed at a crime lab in Atlanta.

My mother said not to worry about it, it was just a meaningless police routine. She was walking away from me, up the stairs. I was standing on the first step, leaning against the mechanical chair when she told me about it. She said she was very tired and was going to take a nap.

There's a good view of the San Francisco skyline from the maze at Grace Cathedral. You can see the Fairmont and the Top of the Mark and Harry Denton's Starlight Room. You can see a sliver of the Bay Bridge. You can even see the Pyramid.

Many tourists come to the terrace at Grace Cathedral to take in this view. At night, it's a popular spot for looking at the stars. Many people walk right over the maze as if it were just a patch of sidewalk, but occasionally a young couple will decide to walk through it. They do so as a lark, not trying very hard to keep to the pattern. When they get to the center, they usually kiss each other, as if they'd just won a doll throwing baseballs at milk bottles.

We celebrated my 16th birthday in my room at Mrs. Darlington's house. My mother sat on my bed and smoked a cigarette. I took the wrapping paper off a new record player and a small set of speakers. "The guy at The Record Bar said you'd probably like those," she said. I thumbed through a stack of records she'd given me tied up in a red bow. "He said if you didn't like them, you could take them back and pick out something else." I plugged everything in and put on a record.

"I have some good news for you," she said. She went to the window and opened it to air out the smoke. It was raining and the air was very cold. "We don't have to take any lie detector tests. The blood they found on your bed was Sherlock's." (Sherlock was our basset hound; he'd died a few months before.) I asked her why they'd wanted *her* to take the test. I thought I was the only one who had to prove I was telling the truth. "It was just a routine. They even wanted to hook Lorraine into that thing," she said. I uncontrollably said "Ha!" The sound of my voice frightened me.

Lorraine, our 50-year-old housekeeper, had found Tom dead when she came in to work that morning. She'd called the police and then called my mother's business partner, Julia. She told Julia there'd been an accident at home and that something was really wrong. Julia wanted to know what the hell had happened, but Lorraine just said, "Tell Mrs. Hiller to come back home. Something's really wrong here." She asked Julia to drive my mother home immediately. She was scared and the police were on their way. It was 8:45 in the morning and Lorraine didn't know that I was upstairs in my bed watching "Captain Kangaroo" until "Donahue" came on at nine.

"They're here," my mother said. I went to the window and saw Patsy Moore getting out of her mother's Mercedes. She was wearing a yellow slicker and L.L. Bean duck shoes. She ran up the stairs, and a moment later Mrs. Darlington's doorbell rang. Patsy's mother saw us at the window and waved. Patsy was in my class, but we weren't friends. I couldn't figure out what she was doing coming to visit, but she seemed happy enough to be there. She'd brought me an album for my birthday—Led Zeppelin IV.

We listened to it and we played LIFE and my mother occasionally brought us more Coca-Cola and birthday cake. I didn't think to tell Patsy the good news about not having to take a polygraph test.

Children take the maze at Grace Cathedral very seriously. They always try their best to stay within the path. They always walk through many times and they never seem to get bored or frustrated. It's as if they simply do not realize that there aren't any real walls or dead ends or cul de sacs. They never seem pleased when they get to the end; they always run back to the beginning to take another pass.

Since my bedroom was at the top of the stairs, roughly 75 ft. from where Tom was shot, the police were somewhat mystified that I hadn't been woken up by the gun blast. "A three-five-seven, hollow-point slug fired from a Smith & Wesson Magnum makes quite a disturbance, son," said one of the homicide detectives. He was fat and very confident, and he made a painful sound when he breathed, like an old dog. "It would've been like a power transformer exploding in the next room over."

I told him, again, that I had been asleep.

After they put Tom on the coroner's gurney, the detective came over to me again, this time carrying Tom's gun in a plastic evidence bag. "Am I going to find your prints on this?" he asked, looking back over his shoulder.

He knew that my prints were on the gun, but I said that they weren't. I was terrified of what might come out of my mouth if I tried to explain. I had gotten into the habit of rifling through Tom's jacket pockets and dresser drawers while he was at work. He took a lot of business trips, and, if I was lucky, I could always find a few deutschemarks or yen or pounds that I could exchange for dollars. I imagine I was the only 15-year-old in Savannah who was obsessed with the current rate of exchange. And I always felt like quite the well-traveled young man, as I'd wait for the woman down at the Trust Company Bank to calculate my take.

Although I had seen the gun holstered in the top drawer of Tom's dresser on many occasions, I had always been scared of touching it. I knew that if I were caught in the petty thievery of foreign monies, I could withstand whatever punishment might result. However, I was certain that if, by accident, I shot a hole in something, I'd be in some seriously deep shit.

About a week before Tom's murder, I got up the nerve to take the pistol out of its holster. I held it in my hand, felt the weight of it, and then immediately put it back. I had been with Tom when he shot at cantaloupes down beneath the Old Talmadge Bridge, so I had seen what the gun could do. It was loud and insanely destructive. I put it back in the precise spot I found it.

I had seen enough movies to know that if I gave the detective a full confession involving embezzlement and gunplay, I would immediately be sent to the electric chair. So I denied having handled the Magnum.

A few days later, the police matched my prints. Further analysis, however, cleared me: it was determined that I had never touched the trigger. Since other people's prints remained on the trigger, the crime lab in Atlanta reasoned that it would have been impossible for me to wipe down my own prints, while leaving the other prints undisturbed.

If you're on the *Midnight* trail and want to get from Mercer House to my house by car, you go two blocks down Taylor, turn right on Barnard and go two blocks to Charlton, turn right, and it's about half a block down on the right. It's directly across from what is now the women's dormitory of the Savannah College of Art and Design, which was the Salvation Army headquarters at the time my stepfather was killed. They'd have band practice there every Saturday morning and we'd always get woken up by tubas and French horns.

My parents split up when I was three, so I have no memories of my father ever living in Savannah. He moved to Philadelphia, and, over the next ten years I only saw him in the summers when he'd cart me and Angus to East Hampton to visit my grandmother.

She had a massive house called The Tide that was about a block from the beach and about a quarter-mile from the Maidstone Club. Angus and I'd always show up that first day in jeans and *Star Wars* T-shirts or something, and she was never pleased with our haircuts. So, on the second day of our visit, we'd hop in her Wagoneer with her dachshund, Rufus, perched on the seat between us, and head down to East Hampton Village for total makeovers.

She'd pick out a stack of polo shirts for us—always a pink one in the bunch—and we'd run the tailors ragged at Mark, Fore & Strike getting suited up with madras jackets, yellow trousers, tennis shorts, and pristine white socks. Then we'd swing over to the barber shop for shearings. By the time we pulled back into Tideline, we'd look more respectable and she was pleased.

My dad played golf every day, taking time out to have a sandwich with us at the beach club at noon. My brother was always running off to Junior Activities, a club-sponsored program designed to wear kids out with swimming and tennis and nature hikes. I never went, because I was shy and very nervous about the whole East Hampton experience.

I'd sit around the house with my grandmother and shell peas or read, but no matter how much she tried to push me into it, there was no way I was going to Junior Activities. She was worried about me, because I was skinny and I'd never eat. She'd have her cooks make anything I wanted. But even though I'd request lamb with mint sauce, or melon balls, or hot dogs, or key lime pie, I rarely touched a bite. She kept some Ring Dings in the breadbox in the kitchen, which I lived on at night after everybody else had gone to bed.

My grandmother sold her house when I was about 18; then

she had a stroke and went crazy for about four years until she died. She was always very sweet and if I hadn't been so socially confused whenever I was in East Hampton, I probably could have enjoyed hanging out with her more.

When she died, she left me some money. And that's what I'm living on these days while I look for a job.

Meanwhile, back home in Savannah, my mom had married Tom Hiller a year and half after putting the kibosh on my father. Like I said, I was only three at the time, but I've gathered enough information over the years to guess as to why she sent my dad packing. Hell, she was only 29 and already she had played the dutiful Savannah Junior League mom bit—parties at the Telfair, picking out upholstery patterns, coordinating cooks and nannies. I suppose that when the seventies came around, she just couldn't bear the thought of another fun decade slipping by while she watched from the sidelines.

I'm sure my father never saw it coming, because he he'd given my mother a Coupe de Ville only days before she told him to pack up his shit and go. I think he'd been happy with the way things were going up until that day. His appliance company was just starting to turn a nice profit, he had a good-loking young wife, a big white house with a covered swimming pool on 44th St., a very advanced sprinkler system for the yard, and two kids that he was entirely devoted to. When my mother served him with the papers, he moved across town to a second-rate condo development called, oddly enough, Tiger Woods. My father was a scratch golfer at the time, but because it was 1970, the irony of his living at Tiger Woods was lost on all of us at the time.

For months, my father fought for custody of me and my brother. But, regardless of the luxurious and straight-shooting life he promised for us kids, every downtown lawyer assured him that, in Savannah, Georgia, taking a mother's boys away from her was like getting weevils away from cotton. So one night he snuck into our garage, "borrowed" my mother's new Cadillac, and moved to Philadelphia. She never pressed charges, she said later, because she had her eye on a black Corvette. That was just before the fiberglass models started showing up in 1971.

Then she met Tom. I hate to boil him down to a sentence, but Tom was essentially a tall, skinny German guy who drank like a fish and liked having a good time. He was a couple of years younger than my mom, and she adored him. He had been born in Holland during the war, his mother having fled to avoid the bombings in her hometown of Bremen. When Tom was 17, he jumped a freighter in Wiesbaden to avoid the draft. After a few wild years in England, he came to the U.S. and settled in Savannah. He said *wie geht's* instead of how's it hanging, called my mother *maus,* and got a *bitte* from me every time he said *danke* for passing the peas. But he tried hard to fit in, like any good ol' Southern boy.

My real father was as sound as a new penny—hard-working, thrifty, and painfully honest. He was very sure of himself and always knew where his life was going. Tom was the opposite, which is probably why my mother got such a kick out of him.

Tom liked to drive fast, especially when he had a Pabst Blue Ribbon tucked snugly in his crotch. He got a few DUIs, but so did everybody else. He also liked to think he was on the verge of hitting the jackpot. I don't remember the details very well, but I know he worked for an import/export company, and I imagine that being European helped him out in a port city like Savannah. Every now and then he'd come racing home and say something like, "Everybody pack, we're going to go live on a houseboat."

Tom and Angus got along very well. When Angus was 17, he began going to the bars where Tom liked to go. Tom and I never seemed to connect. I was always a little kid he had to get parental on—grounding me from going to a Cheap Trick concert, hushing me up when I was being loud and obnoxious. I think he knew how close my mother and I were, so he concentrated his attentions on Angus.

I remember, one afternoon when I was twelve, going down to Phil's Grill on Drayton St. to meet Tom for a sandwich. There was a pool table and after we finished eating Tom asked if I wanted to shoot a game. I said sure, but instead of getting up and grabbing a cue, I asked him to tell me how to play, because I'd never played before. "Pick out a stick," he said. "I'll show you."

But I didn't get up. I wanted him to explain the rules and the principles of pool before I actually took a shot.

"Come on," he said, getting up. "It's easier if I show you."

But still I didn't get up. He was getting frustrated, so Phil stepped over and started telling me how to play.

Tom just sat there and listened. When we finally did play, he slaughtered me. I didn't sink one ball. He was pissed off.

He didn't teach me how to hunt, either, or how to drive, or enjoy a Budweiser, or smoke a Pall Mall. He didn't explain why he liked classical music or why his back hurt all the time. On the other hand, he never hit me or told me to shut the fuck up.

He did do two really great things for me. First, from the time I was a little kid, he let me hang out in the living room when my mother and he entertained. I got to contribute to the conversation, and I was treated like an equal. If I started to cry or got too hyped up on Cokes, he'd send me upstairs. But as long as I acted like an adult, I could hang out.

Secondly, and most importantly, about three months before he was shot, I went out on my first real bender. I was drinking grain alcohol and Kool-Aid over at Carl Franklin's parents' house and I got completely smashed. Somehow I made it home—a survival skill I still have today—but I passed out in our foyer and pissed all over myself and this gorgeous oriental rug I flopped on.

When I woke up the next morning, I didn't remember much, but when I went downstairs and saw all the bath towels on the rug, I sort of put two and two together. My mother never found out, and, later, all Tom said was not to let it happen again. That was pretty cool of him.

After Tom got shot, my father had had enough of this Savannah bullshit, so he and Mom decided it was time for me to go to boarding school. So away I went. □

John Ryan recently decamped to Atlanta. When pressed, he will admit that mazes have deadends and the one in front of Grace Cathedral is actually the Melvin E. Swig Memorial Labyrinth. This was his first time in print.

TRAVELING WITH MOTHER

Katherine Taylor

*T*here are no female gynecologists on my family's medical plan. I didn't see a doctor for six years, because I refused to be examined by one of my father's golf partners. When the dog had cancer, Daddy Taylor flew it from Fresno to UC-Davis for special treatment. When my hair was falling out, he sent me to see a man who told me, "Eat more protein. See a psychiatrist. Your father has a fine swing."

My mother didn't think I needed a psychiatrist. She thought I needed a leisurely mother-daughter drive across America.

That summer, people in the Midwest were dying from the heat. I had never thought about the Midwest except as space on the map between California and New York. That summer, my brother ran an unmarked cop car off the road on the 99 between Stockton and Sacramento and was charged with assault with a deadly weapon. My grandmother was moved out of her house and into The Home. My obese Auntie Petra lost 75 pounds by having a shake for breakfast, a shake for lunch, and a sensible dinner.

My hair came out in clumps. In New York, I was worried and nervous and couldn't concentrate. I had moved there the previous autumn to act and write, but found myself too homesick to do anything but cry and socialize. I threw enormous parties to make myself feel less lonesome, and my neighbors left nasty notes threatening to tear me asunder. A man wearing roller blades molested me on 68th St. I had the screens removed from my windows in case I decided to jump out. Instead, I decided to fly home to California.

In Los Angeles, I called my hairdresser. His name is Armando and I roll the R. I said, "I don't care how busy you are. I have an emergency. My hair is falling out." He said, "First you come, we

cut it all off. Then you stop worrying about whatever you worry."
He rolls his Rs too. Armando knows all my secrets.

In Los Angeles, I met with a producer who had written a part
for me in a film financed by rich Germans. I told him, "I'm not an
actress anymore. I won't prostitute my emotions." Afterwards, I felt
ridiculous. Afterwards, I wondered why I seemed to have no
control over the things that came out of my mouth.

The dog died the day I arrived at my parents' house in
Fresno. The bionic dog, the three-thousand-dollar-chemotherapy
dog. Cancer ate his ears off. Daddy Taylor had brought that mutt-
dog home after he ran over my Dalmatian with his truck on my
tenth birthday. My brothers and I had marked the Dalmatian's
grave in the backyard with a cross and stones. After the mutt-dog
died of cancer, I suggested we bury it in the pet cemetery outside.
Daddy said, "What cemetery?"

I said, "Where you buried Buttons after you smashed her."

He said, "Katherine, I scraped that dog off the driveway and
threw it in the garbage."

I said, "That's against sanitation laws."

My mother agreed. She got wild-eyed and said, "Your father
yells at me when I break the speed limit."

In Fresno, the internist I saw for my hair asked if I were
anorexic. I told him no. He told me I was probably anorexic
whether I realized it or not. His nurse took my blood and stuck
me with a needle three or four times before she found my vein.

The dermatologist looked at my scalp and told me I could be
having a nervous breakdown. I told him I felt too bored to be
having a nervous breakdown. He told me my hair would grow back.

The gynecologist was a family friend. He said, "Your father
has a full head of hair. Your reproductive organs are all in order. I'll
see you at brunch on Sunday."

My mother waited for me outside each doctor's office in her
car listening to news on the radio. When I returned after an
unenlightening examination, she would say, "Those doctors don't
know what they're talking about. You and I need to go see Mount
Rushmore." I told her I had no interest in Mount Rushmore. I
told her I could not possibly enjoy Mount Rushmore while I was
dealing with the trauma of hair loss. I said, "Mother, I have no
time in my life to go driving about with you all summer." She said,
"You'll be sorry when I'm dead."

She called AAA to map out our trip. Mother began to see Mount Rushmore as the promised land, the solution to our suffering. We packed the car full of bottled water and rice cakes. We bought packages of red licorice and vitamin C. Mother bought dozens of Peppermint Patties and locked them in the glove compartment. She told me, "These are for me and you can't eat them. Let me have only one a day." My father bought me books on tape: *The Prince,* Oliver Sacks, the complete works of Kant. I brought along language tapes and thought I might spend the trip learning to speak Hebrew.

The first day we drove as far as Portland, where Mother was upset with the hotel accommodations. She refused to get out of the car at the Holiday Inn. "I don't like the looks of this place," she said. "There's no one out here to help us with our luggage."

I told her, "Mother, this is a road trip. You cannot have fancy room service and concierge on a road trip."

In our room she found a toothpick on the floor and refused to take off her shoes. She wouldn't use the bathroom or touch the phone without a tissue. In the morning we drove seven miles out of Portland to find an IHOP, because Mother craved Cream of Wheat pancakes. She complained about the service. We were bored by downtown Portland and left early for Washington.

We stopped at a gas station where Mother bought an aerosol can full of bleach. "For the toilets," she said.

"You never sit on the toilets anyway," I said.

"You never know what could happen," she warned me.

I didn't know you could buy aerosol cans full of bleach.

At the Seattle Sheraton we changed rooms four times. The first room had an acceptable view, but construction eight floors down that Mother thought might disturb her napping. The second room was quiet, but hadn't as nice a view as the first. The third room was quiet and had a nice view, but wasn't as large as the first or second. When we arrived at the fourth room, Mother told the bellhop, "This will be fine. Leave our bags here and thank you very much." The bellhop refused to leave, suggesting we sit in the room a while and try it out before we made our final decision. Mother waited a moment. We all stood still without speaking. Eventually, she said, "No, this room won't do either. I hear a buzzing sound." I heard no buzzing sound. The woman at the desk said it might have been the air conditioner. The hotel moved us once more and sent up a bottle of champagne.

We ate dinner at a fish restaurant on the pier. Mother disapproved of where we had been seated. "I don't understand why we cannot sit by the window," she said. She protested to the hostess, the waitress, the bus boy, and finally to the manager, who brought us free oysters and told us the window seats were all reserved. I had a martini.

Mother tired quickly of Seattle. She tired of latte and people in combat boots. We tried a trip to British Columbia; we thought we might see the gardens there, but Mother tired of that, too, after bad sandwiches and a quick argument with the people at the desk of the Empress Hotel. Less than 24 hours after our arrival in Seattle, we began heading east.

In the car, Mother insisted on listening to Christmas music. Through Washington and half of Idaho, I slept and Mother sang along to "Feliz Navidad." She drove 100 miles per hour.

At dawn in small-town Idaho, she turned off the freeway. We had driven all night, unable to find a hotel to suit Mother's standards. Idaho smelled like pine trees and wet dirt. I needed to brush my teeth.

"Would you like me to drive?" I asked.

She pulled into an empty IHOP parking lot. "I'm stopping here."

"I'll drive."

"No, we're going to eat here."

"I think they're closed."

"I want my Cream of Wheat pancakes."

"Mother, they're not open."

"We'll just wait here until it's time."

We waited two hours. At precisely 7:30, when the doors to the pancake house should have opened, Mother banged on the locked front door. "Let us in! You have hungry travelers waiting!"

"Don't be so disagreeable," I told her.

"I'm not disagreeable," she said. "They're late opening." She banged again until an ordinary blonde girl in braids and white nurse shoes unlocked the door. "I'm terribly hungry," Mother said to her. "I don't mean to be disagreeable." The blonde girl seated us by the window.

I smiled to cheer myself up and drank a pot of coffee.

Through Montana, Mother slept and I learned Hebrew from tapes. She snored and yelled at me, even while dreaming, if she sensed I was driving faster than 80 miles per hour. Outside of Bozeman, she woke up. "I need a shower," she said. "I need some room service and a bed and toothpaste."

"There's a Best Western in Bozeman."

"No! I cannot tolerate Best Western! No Holiday Inns, no Quality Inns. Let's listen to Rush Limbaugh."

"No," I told her. "Don't you want me to be bilingual?"

"I need some soap." She picked up the cell phone and dialed the American Express travel office. "Hello, this is Elizabeth Taylor. Not that Elizabeth Taylor. I'm in the middle of Montana and where is the closest Four Seasons?"

"You're making my hair fall out."

"How far is that from Montana?" She turned to me, "The closest Four Seasons is in Minnesota." She spoke into the phone, "Well, I can't drive that far! I need something immediately. Thank you very much." She hung up.

"You're *rofef,*" I told her. I had learned that from my tape. It meant "crazy," and I was excited to say it.

"There's a hotel inside Yellowstone Park," she said. "We can turn around and drive there." She navigated from an AAA map.

We arrived at the Yellowstone gate early in the evening. The woman at the entrance booth told us all accommodations within the park were booked and, in order to exit the park before it closed, we would have to drive through without stopping. Mother looked as if the news quite devastated her. I told her not to worry.

We drove through the park slowly, stopping frequently to look at deer, buffalo, hot springs. Mother wasn't impressed. "We've been here 15 minutes, and I haven't seen one bar yet!" she exclaimed. "I'm going to get our money back."

"There are not supposed to be bars in national parks, Mother."

"There are, too. What do you know?! Davy Crockett killed him a bar when he was only three."

The parking lot was full at the Old Faithful Inn. I told Mother to wait outside. I took her credit card and went in to the desk. "Do you have an available room for Elizabeth Taylor?" I imagined myself quite together and beautiful and tried to forget I hadn't showered in two days.

"Pardon?" The receptionist was a college boy whose name tag said, "BOB Oregon."

"Elizabeth Taylor would like to stay at the hotel tonight." I spoke quietly and in my best Hollywood lockjaw.

"Elizabeth Taylor?"

"Yes, thank you." I handed him my mother's credit card.

He read the name on the card and smiled excitedly. "Just a moment." He left and came back a moment later with the manager and a key. "The hotel would be honored to have Ms. Taylor stay with us," the manager said.

"Oh, she'll be very pleased."

I collected my mother outside and shuffled her in past the desk beneath a hooded raincoat.

That night, we ate grilled-cheese sandwiches I bought at the Old Faithful Diner and drank bottled iced tea from the Old Faithful Market. We tried to sleep early, but the beds were hard. We tossed about, each pretending to be asleep.

In the dark, Mother said, "Is your hair falling out?"

"Not so much anymore," I told her. We listened to the awake people in rooms around us. "I'm sleeping," I told her.

She paused a moment, sighing. "Are you sad, really?"

"I'm not sad." I said it as if she had asked a ridiculous question.

"Are you?" She was whispering now.

"Yes, Mother. I am sad. My hair is falling out."

She lingered a moment in the space between speaking and not. "Daddy's friends think you need a psychiatrist."

"They're all bored Fresno doctors. They're afraid of me." I whispered, too, now. "I don't want to be on those drugs."

"I don't, either," she said. "I want your hair to stop falling out."

"Me, too."

"Maybe if you were happy."

"I don't have to always feel like being happy, Mother." I turned to the wall, away from her.

"Neither do I," she whispered, still facing me.

"Those doctors are rofef."

"Daddy's friends are idiots."

"Be quiet, I'm sleeping."

The next day, we drove about Yellowstone. We saw buffalo and moose, deer and volcanic activity. We walked off the paths, and park rangers yelled at us. We were too impatient to wait around for Old Faithful to erupt.

We missed the heat in South Dakota. The day we found Mount Rushmore, the rain came down in hurricane volume.

Tourists took refuge in the gift shop or snack bar. Mother and I stood outside in the lightning, drenched but not cold, staring at the carved faces.

I said, "This isn't as climactic as it was supposed to be."

"I should have seen this a long time ago," she said.

We had an argument just outside of South Dakota. I didn't give her enough warning for a turn-off, and we had to drive an extra 32 miles to get back on our route. She shouted at me with all her might. Neither of us apologized. By the time we arrived in Minneapolis that night, we had not exchanged a word all day. She finally said, "If you don't want to travel with me, I'll leave you off at the airport and you can just fly back to New York by yourself."

"That's fine," I said.

"Roll down the window and ask someone where the airport is."

"Find a gas station."

"Just roll down the window!"

"I hear you, Mother, don't shout."

"I'm taking you to the airport."

"Fine."

"I don't know how we're going to find the airport."

"I'm sure there will be a sign."

We never saw a sign. Instead, we found the Four Seasons and stayed there overnight. Mother's mood improved considerably. We did not change rooms. Still, she refused to take off her socks and told me twice not to sit on the bedspreads. "People have sat on those with their dirty bottoms," she warned me. We ordered hot chocolate and sandwiches from room service.

My hair had stopped falling out by the time my mother left me in New York. The doormen were happy to see her. She tips them every time they open the door. Normally she likes to visit for weeks on end, but this time she didn't stay long.

"I have to go to your little brother's felony hearing."

"He'll be happy to see you there."

She bought me new towels, dusted the bookshelves, and put all my books in alphabetical order.

She continued down the southern route, alone, to California. □

Katherine Taylor lives in Los Angeles. This was her first nonfiction in print and won a Pushcart Prize. E-mail: katetaylor@earthlink.net

ON THE COUCH

Brenda Webster

> No other system of thought in modern times, except the
> great religions, has been adopted by so many people as
> a systematic interpretation of individual behavior.
> Consequently, to those who have no other belief,
> Freudianism sometimes serves as a philosophy of life.
> Alfred Kazin

I was born and brought up to be in psychoanalysis
and, as a result, much of my adult life was spent on the couch.
My family lived on New York's Upper East Side during the rich
yeasty time, filled with new ideas and movements, after World War
II. My father, Wolf Schwabacher, was a prominent entertainment
lawyer. Dorothy Parker and her circle were social acquaintances,
and his clients included playwright Lillian Hellman, said to lie even
when she said "and" or "but," and Erskine Caldwell, whose novel,
Tobacco Road, sparked an obscenity trial that was pure theater. As a
young man, my father had a bohemian side: he was engaged eight
times and once popped up naked from under a table at a Marx
Brothers party. My mother, Ethel Schwabacher, was a protégée of
Arshile Gorky, and after he hanged himself from the rafters of his
barn she became his first biographer. Later she was recognized as
an important Abstract Expressionist painter in her own right, one
of the very few women in the movement. My parents were
idealistic, acculturated Jews, and glamorous (so glamorous that it
was hard not to feel like an ugly duckling born into a family of
swans).

But in addition to her beauty and her passionate love for my
father, my mother brought into the marriage a serious history of
mental instability in her family. Her brother was psychotic and her
mother, while not obviously crazy, was frantic with anxiety and
sought analytic help. My grandmother's need to be propped up
emotionally was a burden to my mother, who became similarly
intrusive and demanding toward me.

By the time I appeared, in 1936, my mother had already had
two analyses, setting up a family pattern of submission to analytic

authority that made me run back to my analyst after every crisis, and kept me there, trying so hard to be "good" that it would be laughable if not for the years of pain and wasted opportunity.

Because I came from such a privileged family, the analysts who surrounded us were only the best. Ruth Mack Brunswick, one of Freud's inner circle and a family friend, analyzed the Wolfman, Freud's famous patient who was supposedly driven mad by the sight of his parents copulating like dogs; Dr. Marianne Kris, Mother's analyst for 30 unconscionable years, analyzed Marilyn Monroe; and my own analyst, Kurt Eissler, a passionately intellectual German Jew who worshipped Freud, was the founder of the Freud Archives. They were the first wave of Freud's disciples to come to New York—many of them refugees from Hitler. They created a powerful Freudian orthodoxy and represented the cream of the American psychoanalytic elite at the height of its power. In a short time, their beliefs permeated American culture and, for as far back as I can remember, my sense of home.

My earliest memories are filtered through my mother's psychoanalytic lens. She liked to recall her favorite incidents from my childhood and delightedly repeated them to me over the years.

My own first memory, when I am five years old, is of her standing by the window in a pale peach silk kimono, covered with exotic birds. Her stomach is flat. She is thin and beautiful again.

"If Grandma loves the baby so much," I tell her, "let's cut him up and send him to her as a present."

My mother doesn't raise an eyebrow. She doesn't take me in her arms and hug me or say she loves me. She tells me it is natural to be jealous, to hate my new brother, even to want to kill him. He is guarded by a white starched nurse. I am never allowed to be alone with him, to hold him.

My next "memory" is of stealing his nursing bottle and running down the long, thin hallway of our New York apartment, the nurse in hot pursuit.

By the age of six, I knew myself as a potential murderer and a convicted thief, an envious and jealous child—destined, like the other women in my family, for the analyst's couch.

My parents' friends, their conversation, their whole way of looking at the world, were steeped in the culture of psychoanalysis. My family's involvement spans 70 years and covers the major part of the movement's history, from its beginning after World War I, the heady days when it was associated with revolution and change, to the time when it became rigid and hostile to new ideas. When I went to my first child therapist at the age of 14, it was the fifties, the golden days of psychoanalysis. Everyone in New York was going or had gone. But it was also a reactionary time, particularly for women, and analysis, at least in my case, fed into the middle-class status quo.

Freud came to America in 1909 on a lecture tour, but it wasn't until after World War I that analysis began to catch on. It got a big boost when the "talking cure" was used with shell-shocked soldiers—it worked better than the previous brutal methods of getting them back to the front. After the war, adventurous bohemians and intellectuals began to try it out. For them, analysis meant sexual liberation, free love, freedom from constraints. At that time, a typical analysis could be over in a matter of months, your fixation located—mother fixation was especially popular—and your psychic energy freed. Men left their wives and went to live in the Village; the first novels about analysis (Lewisohn's *The Island Within* and Floyd Dell's *Moon Calf*) were written. It was a unique moment for a unique group of people.

My mother's family was ripe for a new faith. My maternal grandmother was an Oppenheimer, a member of an old German-Jewish banking family. Mother's family tree stretched back to the 17th century when an ancestor financed the wars of the Christian princes. In more recent incarnations, the Oppenheimers were diamond merchants and theater owners. They were completely assimilated, leaving their children with an ideological vacuum. In succeeding generations family members divided between an attraction to Marxist-Leninism and a belief in psychoanalysis, with a hard core keeping their faith in the sanctity of money.

My mother grew up in Pelham, an exclusive suburb outside New York. The household included nine servants, and the banquets her parents gave were so lavish that she remembered sometimes being sick after them. My grandfather Eugene ate steak and ale for breakfast and took my grandmother to German spas. It was a world of prosperity and seeming well-being. But in the early twenties, my mother's brother James, a brilliant mathematician, became mentally ill.

I didn't learn of my Uncle James's existence until I was in

college. For years, Mother kept silent about her brother. When I was a child, photo albums with his picture were hidden away. My grandmother wasn't allowed to speak his name. His psychosis was a dark secret at the center of our family. Mother was terrified that madness and genius went together, that she was crazy too, that bad blood would be communicated to her family.

James's early symptoms were largely denied by my grandparents. When Mother was in her seventies, she told me that when they were children James had climbed in her bed to molest her. When she ran to her parents for help, she said, she was told to go back to bed and stop making such a silly fuss. They simply didn't believe her.

Ignoring what James did with outsiders was more difficult. He got into trouble—family stories differ as to whether he hurt a fellow student or simply acted increasingly bizarre—and was asked to leave school. By the time James had his first breakdown, my grandfather was dead. The task of getting her brother help fell to my mother, then in her early twenties. Though James went on to get a Ph.D. in engineering from MIT, married, and had a child, subsequent breakdowns led to his being permanently hospitalized.

In the meantime, Mother found an analyst for herself, Dr. Bernard Gluck, who worked as a criminologist at Sing Sing in addition to maintaining a private practice. By 1925, he was treating Grandma as well.

Grandfather would have been appalled by this development. I can't imagine him taking to newfangled ideas. He was a Southern Jew, an archetypal Victorian gentleman, a lawyer by profession, a scholar by inclination. His letters to my grandmother show him tenderly protective of her. As Grandma said later, he felt himself the guardian of the honor of any woman entrusted to him. At the same time, he felt entitled to make the most of his own freedom. For years he had fooled around with the nursemaids, even— according to Mother's reconstruction in her own analysis—getting my mother's nurse pregnant.

After his death, my grandmother was ready for liberation. She had suffered during her husband's long bout with stomach cancer and had resented his philandering. Maybe Grandma thought she could experiment and live a little. Her aging parents and her authoritarian brother, George, might have restrained her, but she had already distanced herself from her family by marrying a scholar/lawyer and by letting my mother study art. After my grandfather's death, she threw herself into a variety of beliefs and practices: mesmerism, spiritualism, Christian Science, and psychoanalysis.

Mother never was explicit about her reasons for going into therapy with Dr. Gluck—her therapy began before Grandma's—but there are plenty of clues. In her journal, Mother describes at length her jealous possessiveness as a child and adolescent and her desire to be her mother's only love. Grandma's letters to their shared therapist show her, on her part, closely watching Mother's moods, alternately affectionate and critical, as when she notes that "Ethel had a crying spell. And induced me to comfort her with lavish praise. Lollipops! I suppose it's hard to teach a child brought up on superlatives, as Ethel has been, to accept sincerity."

Just as my mother later worried about the consequences of my sexuality, Grandma confided to the doctor her fears that Mother's freedom and boldness with men, though admirably sincere, would cost her her reputation. Most striking, however, is Grandma's actual meddling. She wrote to one of Mother's beaus, rebuking him for not calling her daughter when his boat docked: "No time? Piffle!" and ending, "Telephone her but don't mention me! Or write me fully and truthfully." Ignoring all reasonable boundaries, she even went so far as to tell Mother graphic details of her affair with a young lover, John McKay. To Dr. Gluck, she reported: "Oh, I know you will reproach me...but have you ever cared enough for me to know what I feel?" My grandmother combined manipulation, coyness, and genuine distress into a lethal cocktail.

It was perhaps inevitable that Mother would think of Dr. Gluck as a sort of savior and fall desperately in love with him—it was the first example of the addictive dependency that marked her later frantic clinging to Dr. Marianne Kris during my own therapy-ridden adolescence. Dr. Gluck, Mother said, reminded her overpoweringly of her lost father. He had her father's square jaw and smoked a pipe. Just the smell of his tobacco was enough to make her feel faint.

My mother loved her father passionately. She was sickly as a child, and he coddled her, keeping her out of school until she was eight because of stomach aches. (In her later analysis, she wondered if this was the first of her neurotic attempts to get attention through illness.) Surviving letters show my grandfather, when away on business, writing to his "delicate child" with the same protective tenderness he showed to his wife. After he died, Mother sculpted a monument for him. He had been extremely patriotic, but despite his eagerness had been too old to serve in the war. Interestingly, Mother's monument to him, a series of powerful bas-reliefs, has a feminist antiwar cast. There is a woman holding a dead child, a group of women waiting, workers—all people behind the scenes. A photo shows Mother smoking a cigarette and studying this work, her velvet blouse showing between the open folds of her smock.

Mother wanted Gluck to become her lover. Given the weight of emotional baggage she brought to the request, his refusal was devastating, and she made the first of a long series of suicide attempts. Sketchbooks from late in her life contain melodramatic cartoons of herself lying on a couch in Gluck's office. Balloons over her head show that she is thinking of murdering his wife and two children. Other cartoon bubbles show bottles marked CYANIDE, or show him escaping on a plane while she is lying helpless on the floor. In one of her rare efforts to make humor out of pain, Mother nicknamed herself not Oedipus but Oedipet.

During her therapy, Mother expressed her love for Dr. Gluck by sculpting a bust of him. A photo taken by Grandma shows him as a good-looking middle-aged man, with a serious but kindly expression. Mother tried to keep on with the bust even after Gluck left the country, but couldn't stand the grief caused by contact with his image. Eventually, even modeling the clay reminded her of how she wanted to touch his skin. If her drive to be an artist hadn't been so strong, she might have given up altogether at this point, but she didn't—she switched to painting. Love and loss became the major subjects of her work, and suicide her way of coping with the distress of separation.

Dr. Gluck's treatment of both mother and daughter would be unthinkable today. It certainly inflamed their rivalry, and probably exacerbated Mother's instability. When Dr. Gluck took flight to Vienna for some re-analysis, his final advice to my grandmother was to put Mother in a sanitarium.

When my mother died, I found a small notebook, which I call "the rage notebook." In it there is a fantasy of cutting off John McKay's penis and shoving it down her mother's throat—choking her with her own lust. (To me this is a perfect example of how classical Freudian analysis failed to help, probably even made things worse for my mother.) She was writing this when she was nearly 80 years old, after 30 years of treatment, five days a week, still trying without success to recover some original childhood trauma, cranking out fantasies like this to oblige her analyst, stubbornly dwelling on the past.

After her suicide attempt, my mother went to Vienna, where she began analysis with Helene Deutsch, one of Freud's early followers, whose consuming interest was female masochism. Feminists have faulted Deutsch for this emphasis, without noticing that in her work she suggests that women fight a tendency to be masochistic by building up their "ego interests," their work and competence. Mother told me later that Deutsch had encouraged her painting and even found her a teacher. She said that Deutsch also cured her of her sadistic impulses toward men and enabled her to marry my father.

Deutsch's role in my mother's life didn't stop when the analysis ended. She presided over my conception like a fairy godmother. I might not have been born at all if she hadn't approved of my father when Mother took him to meet her in Boston. Though Deutsch's comment was a surprisingly lukewarm "He's all right," apparently it sufficed.

It's hard to imagine my dashing, headstrong father submitting to an audition by his fiancée's analyst. I see him arguing a case in court much as he looked in photos of that time, his dark hair smoothed down, wearing a perfectly tailored suit, gold cufflinks showing when he raised his hand to emphasize a point. People called him "the little Napoleon"—short, cocky, exuding vitality and confidence. What would have made him submit to approval this way, as though he were a horse at market?

Well, he was in love, and though he certainly hadn't been analyzed, several of his show-biz clients had tried analysis and he probably thought of himself as a very free young man. In his most famous cases, the ones he had tried just before he met Mother, he fought for the right of artists to treat sexual matters usually considered taboo. When he defended Erskine Caldwell against obscenity charges, he compared the secretary of the Society for the Suppression of Vice, John Sumner, to Hitler. When Lillian Hellman's play *The Children's Hour* was banned because of lesbian subject matter, he countersued the city of Boston for libel. In principle, he would have approved of psychoanalysis's fight against sexual repression. In practice, as a father and husband, he had something of a double standard.

In any case, going with my mother to see Deutsch set the tone for a marriage in which the emotional side of things, for better or worse, was left to her. The emotional side of things included my brother and me. She became the one to explain us to Father, and Father to us, and us to each other.

When I was born the following year, Mother took me to see Deutsch, too.

"This is your greatest creation," Deutsch told Mother warmly, moving my blanket away from my face. The signal was clear: Motherhood is woman's highest goal. Besides, according to psychoanalytic theory, having a baby is the only way a woman can get over the horror of not having a penis.

Photos show my mother, ravishingly beautiful, chiffon scarf wound around her head, holding me naked on her lap, her fingers making indentations in my baby fat. But she was hardly present. Once or twice a day she would come into the room that I shared with my German nurse and ask how things were going. The rest of the time, she painted in the corner of the living room where her easel stood.

My nurses never stayed long enough for me to have a clear memory of any of them. They were always doing something wrong and being fired by Mother. My German nurse was caught tying me to my high chair so she could force-feed me like a goose. My next nurse, an Irish girl, was fired because she let me fall off a slide in the park in 1938, when I was one-and-a-half. I had a terrible concussion and was put to bed for 6 weeks. I remember Mother washing my hair in the bathtub (probably the nurse's day off). In my memory she is huge, looming over me like a giantess; my head suddenly hurts terribly and I scream. The giantess's eyes widen with alarm; I see her huge mouth open. Mother later told me I'd had convulsions and she had to force a spoon between my teeth to keep me from biting through my tongue. When I went to my first therapist at 14, Mother repeated the story because she thought it might have been traumatic. She also told me, proud of her acumen, that when I was finally allowed out of bed, I thought I'd forgotten how to walk. I still have nightmares of falling.

The therapist I had in my twenties remarked that perhaps my mother wished she'd spent more time with me and less with her painting. I don't think she did wish that, at least not when I was a little girl. She was probably right not to. No matter what she did, she would have been an anxious, temperamental mother. She was afraid of babies—didn't know what to do with them. Her anxiety would probably have been worse if she had had to limit her painting.

The one thing she insisted on when she married, she told me as an adolescent, was that she should be able to continue her work. In other respects, marriage changed her. She began to act more like a traditional woman. The change started even before marriage. I remember her telling me that when she visited my father at his farm in New Jersey, she was careful not to give in to his pressure to sleep with him. "I knew," she said, "he wouldn't like that." She had been metamorphosed into a virgin.

Photographs taken before her marriage show her looking seductive in a turban, smoking a cigarette with a long holder, or in velvet lounging against a wall. Photographs taken after her marriage show her to be still beautiful, but she is dressed in white painter's slacks and shirt, or after my brother's birth, in sensible tweeds, her hair wound up in a long coil behind her head. Even though it gave her headaches, my father wouldn't let her cut her hair. She was— as my father gently pointed out to me when I was slovenly or too loud—a lady.

And what about me? A portrait by the photographer Sylvia Swami, who specialized in the children of the rich, shows me with my glossy light-brown hair cascading over one eye like a miniature movie star. Though never as glamorous as my mother, I was a pretty child with dark, expressive eyes and a full, sensual mouth— features which undoubtedly prompted the photographer's pose.

I was the perfect Freudian child: I worshiped my father. Everything about him was wonderful, from his slicked-back hair to the shiny black shoes he wore when he went out to parties. Unlike my mother, he was very physical with me, holding me on his lap, swinging me, letting me crawl into bed with him on Sundays and snuggle while he read the funnies. When I was about three, my mother explained to me in a general way that babies come from seeds, and from then on I desperately wanted my father to give me one so I could grow a baby.

Here is an early image: It's my third birthday. My father comes into my room with a big package. "For you, Bonnie," he says, crouching down next to me, so he can enjoy my happiness when

I open it. "For you, princess." I lean against him for a moment smelling his skin, which has the odor of spice cookies. Then I pull off the shiny gold paper. The doll is thin and stiff with black wavy hair and has a suitcase full of beautiful dresses. I take one look and burst into tears. "But, Daddy, she's too old" is all I can manage to say.

The next doll he gave me was better. She looked like a real baby and had a newly invented, rubbery skin that you could wash, and, if you put water in her mouth, she peed out of a small hole in her behind. A photo shows me happily naked, trying to give her a ride on the back of our Great Dane, Sero. Later, I'm told, I approached my father more directly and asked him straight out to deposit a seed somewhere—the bathroom struck me as an appropriate place—where I could gather it up and grow it. I even pointed to the exact place on the white tiled floor where I wanted him to put it. I didn't want it too near the toilet, because it might slip in and be flushed away.

"Here," I said. "Right here. See."

"He only laughed," my mother told me years later, "and I had to smile myself. It was so classic."

There are things that Freud was dead right about and one of them is the Family Romance. In 1941, the summer before my brother was born, I remember digging in the sand with my father. I am wearing only trunks, not the silly-looking tops the older girls wear. My hair is short. My father and I are pals. We work together from opposite ends of a big tunnel. My arm is in it up to my shoulder. My mother wouldn't be able to do this, I think. She can barely bend over now. She waddles when she walks and has to rest a lot. I feel I am gaining in my father's affections. In a photo from that summer, I have a newly confident, come-hither smile. I am slender and deeply tan. The slightly androgynous look has done nothing to diminish my allure.

My flirtatiousness attracted a host of little boys—one in particular, Johnnie, who had dark shining eyes and was always showing me how far he could jump. My father was annoyed by Johnnie's presence, digging or leaping around me. Mother pacified him by telling him—in her capacity as interpreter—how much the stocky, energetic little male resembled him. She told him he should look on it as a form of flattery.

My mother, not I, got the coveted baby. He was chubby and ugly, cried in a loud voice, and had a tiny sausage of flesh hanging down in front. Seeing my father look at my brother with the pride and affection that had up to then been all mine was shocking. Especially painful was the expression of self-congratulation and fellowship that I noticed without quite understanding. My father had produced an heir, and I had suffered my first betrayal.

The analytic community in New York had grown during the thirties and forties with an influx of refugees from Nazi Europe, among them Heinz Hartmann, Rudolph Loewenstein, the Krises, Berta Bornstein, Gustav Bychowski. Lawrence Kubie, a wealthy analyst friend of ours, helped many of the immigrants set up new homes and practices. Most of them, like Drs. Kris and Eissler, lived on Central Park West. Some, like Gustav Bychowski, lived on the Upper East Side. The places they lived in coincided with the boundaries of my childhood world, whereas most of the artists lived in the Village.

Yet there was a strong connection between analysts and artists. The artists went into analysis for relief of their problems—and many wrote novels or scripts using Freudian themes—while the analysts probed the mysteries of the creative process. Kubie wrote a book on genius and analyzed Moss Hart. Eissler, also obsessed with genius, wrote on Goethe and Hamlet. Gregory Zilboorg analyzed Elia Kazan, father of one of my friends from school. Marianne Kris analyzed Marilyn Monroe. Even Mother's interminable analysis was written about for the *Psychoanalytic Quarterly* by a literary critic, Jeffrey Berman, who felt that despite the 30-year sequel with Kris, Deutsch's earlier work with Mother was a great success.

I knew many of the analysts through their children. One of my best friends was Bychowski's daughter Monica, an intense black-eyed girl who arrived at Dalton School from Poland during the war. She told us how her father, not believing in the danger, practically had to be dragged out of Poland at the last minute by her mother. They lost everything—their beautiful house, his art collection, books. I spent a great deal of time at their Fifth Avenue apartment overlooking Central Park, and through Monica met her best friend, Lizzie Loewenstein, analyst Rudolph Lowenstein's daughter. When I was a freshman in high school, I dated Heinz Hartmann's son Ernest, who took me to *Kiss Me, Kate* and then tried to kiss me, fleeing in panic after we bumped noses. He was terribly Germanic, clicking his heels and bowing over my hand, and, even then, clearly an intellectual. I only saw Tony Kris, Dr. Kris's son, once, but I thought he was dreamy and was dying to go out with him.

I wasn't aware when I visited their children that Loewenstein, Hartmann, and Kris were responsible for an important new development in psychoanalysis, ego psychology, which emphasized the ego's ability to master reality. This was a shift away from earlier ideas of freeing the instincts. Given what had just happened in Nazi Germany, perhaps too much permissiveness was dangerous. By that time, failures were also being reported in the area of permissive child rearing. It appeared that permitting expression of violent hostility, such as death wishes against a sibling, vastly increased a child's insecurity. The conclusion of many analysts, including Anna Freud, was that people needed some protection against the force of their drives.

It's one of the ironies of my situation that the famous child analyst whom Muriel Gardiner suggested for me, Berta Bornstein, acted as if she had never heard of ego mastery and continually encouraged me to express my sexuality. Bornstein was Muriel's closest friend among the analysts. It was she who had warned Muriel to get out of Vienna after the Anschluss, and Muriel referred to her affectionately as Bertele. However, Bornstein had seen Muriel's daugher Connie as a very young child—Muriel was worrying about raising her without a father—and Connie hadn't liked her.

Berta Bornstein was certainly one of the scariest-looking people I'd ever seen. She was short and dumpy with wispy gray hair and a red birthmark covering a large part of one side of her face. Besides this, she had a nervous mannerism of putting a cold glass or something else to her forehead as though she were suffering from terrible headaches.

In the first sessions, Bornstein made me lie down on her couch and free-associate. All I could think of was how ugly her birthmark was and whether her husband, if she had one, could possibly sleep with her. When I wasn't thinking that, I would compulsively imagine sucking a penis. I would have died rather than tell her either of these thoughts, so eventually she let me sit up, and asked me about my masturbation fantasies. "Don't worry," she told me, when I said I'd never masturbated, "you did it just like everyone else. Only you feel too guilty to admit it." When I told her about riding bareback, she almost jumped out of her seat. "But of course that's how you got gratification," she said. Eventually she confined herself to giving me practical advice. She wanted to have me fitted with a diaphragm, but she was afraid that if she did I'd tell everyone at school. Certainly her talk about masturbation and her encouragement to get a diaphragm only confused me. Though I clearly needed attention and nurturing more than sex, I also needed to develop some independence. The fact that my sexuality was being managed by my shrink only made me feel more like a baby.

Bornstein might profitably have asked me how I felt having a mother who, though she spoke brilliantly about literature, couldn't talk to me about my feelings and had to dump me on a stranger. If my father had lived, we would certainly have fought over sex, but not only was my mother in such fragile shape, she was also so damned sympathetic. I couldn't get any distance, not to speak of mounting a rebellion.

This may not be fair. Perhaps Bornstein saw that I was going to "do it" and simply wanted to make sure that at least I didn't get pregnant. Or maybe she didn't want me getting married just so I could have sex. She was quite scornful of my first serious boyfriend, Ed, saying, at one point, "You're not really thinking of marrying that little bourgeois, are you?"

Initially, Mother may have thought Bornstein would help control me, while she, Mother, remained my confidante. Muriel was enlisted to help, too. Mother asked her to keep an eye on us when we visited the farm. Of course, we spent our time kissing. Once, holed up in my old nursery, Ed was emboldened to strip. It was my first experience of feeling his naked body. I was shocked and frightened by all the hair: he was covered with fur. At first, it was all I could do to lie next to him. At about one in the morning, having fallen asleep, I heard Muriel at the window.

"It's very late," she said—I heard her muffled laugh—"and I promised your mother I'd see that you behaved."

Though I certainly wasn't ready to settle down at fifteen, sexual exploration had a momentum of its own. My memories are round and bright, full of sensations.

Eddie had a pair of cream-colored slacks, and I was riveted to the way the fabric draped over his erection, the way it rose, the exact degree it stood out from his body. I began to think of it as a little homunculus with a personality of its own. We named it Melvin and I have a photo of Ed in a tight bathing suit, with "love from Melvin and Ed" written in his best calligraphy at the bottom. The beauty of my flirtation with Melvin was that it was unconsummated: the partial knowing, the gradual disrobing, was delicious.

I remember one late afternoon in the backseat of the car. Bill was driving me and Ed and Mother back from Dr. Kris's country place. Under cover of a lap robe I was stroking Eddie's stomach as we sped along, and he suddenly sucked in his breath, letting me slip my hand beneath his belt. So this is really it, I marveled, feeling Melvin's heat.

Another time, at a beach outing with Mom and Chris, after we'd kissed for hours in the water, Ed lingered in the dressing room. I went back to see what was keeping him.

"It got frozen stiff," he whispered through a crack in the gray boards. "I can't get it down." He'd been trying to jerk off.

Through a knothole, I could see him move his hand with its silver ID bracelet confidently up and down. I was fascinated by his casual mastery of his sexual organ. I wasn't even sure what mine consisted of (or where they were). Every morning at school, my friends Monica and Sylvia sat in the hall and swapped information. When I told them that Ed and I couldn't find my clitoris, that it mainly hurt down there, Monica looked at me quizzically. "Gosh. I found mine at six, climbing trees," she said. "Maybe you should read *Ideal Marriage*; it has illustrations."

"Forget about it," Sylvia advised. "Just play with his dick."

"She knows you like ze penis," Bornstein said when I related this conversation to her. "That's good. Very unneurotic." From a Freudian viewpoint, it meant I was moving toward the genital sex that was considered the sign of maturity. But from the beginning of my relations with Ed, it was the idea of sex that interested me more than doing it. I liked talking about it, trying it out as fiction. Anybody but my analyst would have seen I wasn't ready for the real thing. □

Brenda Webster, the president of Pen American Center West, lives in Berkeley. These selections became part of her memoir, The Last Good Freudian *(Holmes & Meier Publishing, New York). Her new novel,* The Beheading Game, *will be published next year by Wings Press, San Antonio, TX. E-mail: websterbrenda1@aol.com*

THE SHADOW OF THE BIG MADRONE

Philip Levine

My first night in California I spent in a motel in Squaw Valley; it was late summer of 1957, and the place was being developed for the coming winter Olympics, but in August it was all but deserted. I was alone, having left my wife and two sons in Boulder with my mother-in-law while I came ahead to find a place to live. I'd come down with some sort of flu the day before and had stopped just before sunset west of Salt Lake City. I'd been seeing things on the road all day, things that weren't there, flying cats and dark birds who disappeared in the shadows; these creatures were beginning to spook me, but it wasn't until I'd stopped for an ice cream cone that I realized I hadn't eaten all day and had no appetite. Very strange. The guy who made the cone for me said the usual, "Hot enough for you?" and as I nodded it struck me that I wasn't sweating at all, but everyone else at the roadside stand seemed stricken by the heat. My forehead was burning, so I drove on to the first motel and stopped. All I had with me were a few aspirins and some antihistamine pills, which I took. Before dark I was asleep. When I wakened some hours later. I was so drenched with sweat I had to move to the other bed. At 5 A.M. I wakened again and got on the road ahead of the truckers and crossed the Great Salt Flats before the day's heat came on. At a filling station in eastern Nevada I asked the man at the pumps what the speed limit was. I watched his eyes behind his sunglasses move across the hood of my teal-green '54 Ford two-door. "I don't think you got to worry," he said. Before dark I'd climbed my third mountain range in as many days. Never before had I seen such dramatic landscapes. In Michigan anything taller than a Cadillac is considered a hill.

Taking a small radio into my Squaw Valley motel room I still felt light-headed and slightly high on nothing. I didn't know if it were due to the altitude or the previous day's fever. I lay out on one of the beds and listened to the most amazing radio program I'd ever heard, on a station called KPFA in Berkeley, which was hours west of me. The program consisted of one man with an extraordinarily affected and ponderous professorial voice reminiscing on the famous people he'd known personally. His articularity and the range of his associations dazzled me: Gertrude Stein, Jung, Robinson Jeffers, Isaac Bashevis Singer, Tu Fu. When the program ended, I discovered it had been Kenneth Rexroth. A true poet on the radio! What a rich world I'd stumbled into. I was so excited I had trouble sleeping that night and once again rose and dressed in the dark. By noon I'd crossed the Bay Bridge into San Francisco singing "I Cover the Waterfront" in my glorious baritone that fortunately no one heard. Ahead of schedule, I stopped at a diner for coffee and directions and was amazed by the graciousness of the counterman, who drew me a map all the way to Los Altos on the peninsula south of the city. There I would find the home of Yvor Winters, who had generously offered to put me up until I found a place to live.

That spring I'd received a short terse letter from Winters informing me that he'd chosen me to receive a Stanford Writing Fellowship. This was a great relief for my wife and me; our second son had come down with a childhood form of asthma and we were advised to seek a more gentle climate than that offered by the Midwest. For two years I'd been teaching technical writing in the Engineering College at the University of Iowa as well as one course each semester in Greek and Biblical literature, and my first teaching had left me with very little time for my own writing. It was the first job I'd had in years that left my hands clean, and I'd begun to wonder if I could both live on my wits and write my poetry, for I'd written much more while I was doing unskilled work in Detroit.

Winters' home on Portola Road was surrounded by a high redwood fence. A brief notice on the gate warned that there were dangerous dogs within; one was advised to use caution and enter at risk. I advanced gingerly. The door was answered by a tall, spare woman whom I'd interrupted at household chores; I took her to be the maid. When I explained who I was and why I'd come, she gave me a wonderfully open and welcoming smile and asked me to be seated. In contrast to her strong, dark features her voice was faint and barely audible, her hair drawn back and largely hidden under a flowered scarf. I recalled a little magical poem by Winters' wife, the poet and novelist Janet Lewis, which depicted the slow movements of a cleaning woman, and I wondered if, like "some Elsie" of William Carlos Williams' famous poem, she were the same maid grown to womanhood.

> Girl Help
>   Janet Lewis
>
> Mild and slow and young,
> She moves about the room,
> And stirs the summer dust
> With her wide broom.
>
> In the warm, lofted air,
> Soft lips together pressed,
> Soft wispy hair
> She stops to rest.
>
> And stops to breathe,
> Amid the summer hum,
> The great white lilac bloom
> Scented with days to come.

Seated, waiting for the arrival of my mentor-to-be, I finally figured out that the woman had told me he was not home. As the time passed slowly I could only hope he would not be too long in returning. I noticed a photo of Winters on a bookshelf behind the television set. (Did Yvor Winters actually watch television?) He had aged considerably since the famous photo of the severe young poet I'd seen in various anthologies; it presented a grim bespectacled fellow in shirt, tie, and leather jacket who seemed in the throes of some terrifying moral problem. This man was actually smiling, perhaps caught off guard, and the woman who stood at his side in the photo was this very housekeeper, who I realized must be Janet Lewis.

The older man I soon met rarely smiled, but for reasons I cannot explain I felt even on that first Friday afternoon that there were stores of affection in him that went unexpressed. He arrived in, of all things, a tiny red English sports car, and directing his gaze steadily into my eyes introduced himself; before I could take my seat again he explained that the chair I'd been using was his and he directed me to another alongside it, and thus we sat side by side conducting one of the most awkward conversations I'd ever been a part of, but Winters was very good at silence. Minutes passed while he stoked and puffed on his pipe; occasionally he would issue forth a heavy sigh. He seemed not in the least curious about me. I noticed that every few minutes he stared into a mirror that gave him a view of the front door to the house, which was behind him and to his left. Apparently he had never heard Satchel Paige's dictum, "Don't look back, something might be gaining on you," or if he had he'd discounted it. Much to my surprise at five that afternoon Winters turned on the TV set to watch a rerun of a Robin Hood serial. "Pay close attention," he said, "it may improve your accent." So he was not entirely without a sense of humor.

That very night I learned he was a devotee of prize fighting. He later assured me that prize fighters and poets had one central thing in common: pride in their abilities. A fighter who doesn't think he can beat everyone in the world is no good to anyone, he told me once, and a decent poet has the same confidence. I too was a boxing fan, and this brought us together at least once a week to watch the Friday night fights, on which we usually bet when there was a difference of opinion. In my incredibly short career as a boxer I'd learned considerably more about the art than Winters had. I had quit after my marvelous coach, Nate Colman, had advised me in one pithy sentence regarding my chief strength. "Your ability to take a punch," Nate had said, after watching me get whacked about by a mediocre light-heavy, "is worthless if that's all you're doing." Would that most literary criticism went so directly to the point. I never lost a bet to Winters, though the largest I ever won was a quarter. I wouldn't say he was not a gambling man, for he'd taken an enormous gamble on his talent as both critic and poet.

Like many Californians of that era, Winters was a hater of some actual or imagined Eastern fight establishment that had managed to keep deserving West Coast fighters permanently from glory. He especially hated Floyd Patterson and his manager Cus D'Amato, neither of whom was part of any fight establishment; they had refused to give the new California hope, Eddie Machen, a shot at the heavyweight title. (When Machen finally got his big chance, he was knocked out in the first round by Ingemar Johansson, and thus it was the Swede who was given the opportunity to dethrone Patterson.) I soon came to realize that Winters felt about prize fighting exactly as he did about poetry: both were rigged by some all-powerful and invisible Eastern conspiracy, and he and his favorites would have to wait on the outside pending some miracle. For all I knew then, he was correct on both counts.

One thing was sure: he knew a lot more about poetry than he did about prize fighting. I later learned that he'd come to boxing as a young man living in what he called "the coal camps" of New Mexico, where he had gone to live on doctor's orders to combat a case of TB he'd come down with in his early twenties in his native Chicago. In order to support himself he'd taught high school in New Mexico, and it was there he'd had to learn the fine art of boxing so as to enforce discipline in his unruly students. It was impossible for me to guess what Winters had looked like as a young man; at 57, when I first met him, he had a thick sturdy body, one that in no way resembled that of the classic fighter. His shoulders were narrow, his beam broad, his arms short, and only the thickness of his neck suggested a fighting past. On Saturday nights, he told me one afternoon as we waited for the fights to come on TV, there had been public fights on the streets of his town in New Mexico; anyone and everyone was welcome to participate, and for weeks he had hoped to take part. But even then he was no fool: not being a large man nor one carrying the heavy muscles of a miner, he needed to learn the finer points of the boxer's craft, and they had been taught to him by an old ex-pro. Laying his pipe aside, he stood in his slippered feet, and showed me how his coach schooled him in the use of both right hand and left hand. He then faced an imaginary foe and pumped both hands forward and back like someone aping the movements of a cross-country skier. "Like this!" he exclaimed.

"I never lost a fight."

"You were lucky," I said.

"After the first fight I had no more trouble from those big miners' sons in my classes."

"You were lucky," I repeated.

What did I mean by that, he wanted to know. I explained that the stance he'd taken was the worst possible one to assume if you were serious about not getting hurt. "Really," he said, "how should I stand?"

I'd known him some months when this exchange took place, and never before had I gotten his attention with such intensity.

"Show me what I'm doing wrong," he said. I began a modified imitation of the first lesson my old coach had given me; modified because Nate always concluded that lesson with a little passage he entitled "Who's boss," in which he'd pin back the ears of the fledgling by delivering a variety of punches that showed the student how little he knew. I had no confidence in what might happen if I were to manhandle Winters, so I merely demonstrated to him that he could neither move backwards nor forwards with any speed and that if I were to shove him he would land on the seat of his pants. I took the proper stance and demonstrated in slow motion that while I could easily reach him with my left hand, he was more than a foot short of me and his entire body was open to my punching whereas most of mine was distant and guarded. He nodded slowly taking it all in. I then arranged him in my stance, left arm and left leg forward, and then showed him how easily he could move forward or back, right leg following left forward, left following right backwards. "That's the way Joe Louis always moved forwards," I said. Winters asked how he moved back. "He never needed to move back."

"That was a counterpunch," Winters said and smiled. He was enjoying this, so I went on to explain that Ray Robinson didn't always follow those basics because he was so gifted he could improvise, he could square up or cross his legs or punch off one foot because he could get away with anything. Winters watched as I demonstrated these moves. "Louis looked like a natural," I said, "but he was someone who mastered the basics so well they looked natural. A Sugar Ray or a Wallace Stevens comes along once in a century." Winters was taking it all in, his cheeks flushed, his mouth loose and relaxed, his eyes wide. He asked me how I'd learned all this, and I told him about my great coach, who had once been the amateur middleweight champion of the U.S. and was in training for the Olympics when World War II put them on hold for twelve years. "He was a lot like you, Mr. Winters," I said. (Yvor Winters, whose first name was actually Arthur, never encouraged me to call him anything except Mr. Winters, and I was comfortable with that name.) "Nate was a purist," I went on, "he believed in the art of boxing and at the same time he thought it was ugly to punch another person for money, so he rejected professional boxing and instead spent his evenings giving free lessons to kids like me." Once again we were seated, and I went on to describe encounters in the gym in which Nate had easily bested professionals, on one occasion a middleweight contender who'd given Graziano a tough fight. Nate had him down in less than a round; he did it all with body punches.

"Body punches?" Winters said.

"Yeah, they would wear light bag gloves; it was a real fight. The guy wanted to hurt Nate, he wanted to take the mastery of the gym away from him. Nate couldn't let him do that, so they went at it with the bag gloves, and Nate didn't want to break a hand on this guy's big hard head, so he destroyed him on the body." On previous occasions Winters had narrated some of the great West Coast matches, and I had merely listened. Now it was my turn to discover what an intense listener Winters could be. What I spoke of I'd actually seen, whereas the fights Winters related were part of a general mythology that passed from one man to another, few of whom had witnessed the events. When I'd finished, Winters nodded. "You must be one hell of a fighter yourself."

"No," I said, "I stunk."

"You're being modest."

"No, I'm serious. My balance is mediocre and my hands aren't nearly quick enough. I fought light-heavies with quicker hands than mine. I'm strong and durable, and that may make it in poetry, but it's no good in fighting."

Winters nodded, convinced.

At the first meeting of Yvor Winters' graduate writing class there were three students. One was the other fellow chosen by Winters; he was an attractive young man named Francis Fike who I had learned was an ordained minister, though he lacked a congregation to minister to. The third was a young poet from Philadelphia who had come to Stanford on a Woodrow Wilson Fellowship. It struck me that without fellowships Winters would have no students at all. Winters had told the third poet to bring a selection of his poems so that the professor could decide if he should be admitted to the class. In Winters' place I would have grabbed at any warm body, but it was immediately clear he felt otherwise. We three students sat in silence while the old man, his face closed in a determined scowl, read slowly through the sheaf of poems. After some minutes of this Winters looked up and said, "This line doesn't make any sense."

The young poet, a short, dark-haired, neatly dressed fellow in slacks and sweater, seemed far less distressed than I would have been. "What line is that?" he said.

Winters read in his deep sonorous voice, "'At dawn the young grass wakens on darkened legs.' The movement wouldn't be terrible if it were in the proper context of blank verse. As it is it's set in a passage that's not verse at all and it doesn't mean a thing."

"It is poetry," the young man said, "and it has a meaning."

Winters leaned back in his leather swivel chair behind the great oak desk. He was within easy reach of his poetry library which was arranged alphabetically in bookshelves that reached to the ceiling on three of the walls. The three of us sat on metal chairs across the desk from him. It was a small, cozy room with a single window that opened directly on to the quad. It was easy to see why Winters spent so much time here. "Perhaps there are three lines of verse in this one, three of..." and he scanned the page "...three out of twenty-five or so, but this line means nothing." The young man began to answer, but Winters held up a palm. "I've lived with grass. I've grown every kind of grass you can possibly think of, even Jimson weed, and I can assure you the line means nothing."

The young man repeated himself. "It's poetry, and it has a meaning."

"If it's poetry," Winters said, puffing on his pipe and shrugging his shoulders, "it's very bad poetry, but in any case it has no meaning."

The young man sat calmly and in silence. He appeared to be totally above the fray, confident in his abilities as a poet and in no way distressed by Winters' display of candor or bad manners. I thought to myself that if every class went this way I was in for one tortured year. I hoped I was just seeing Winters' version of what Nate Coleman had called his "who's boss" initial lesson. A few minutes passed. Outside I could see the students passing quietly down the bricked paths of the quad. Everyone seemed to behave as though they were in a library. I'd had no idea the name Stanford resonated so potently. Winters broke the silence. "You insist it has a meaning. Fair enough. Tell me what it is."

The young fellow rose from his chair, nodded first at Fike, then at me, and at last addressed Winters in a quiet, unruffled voice, reaching first across the desk to take the sheaf of poems from Winters' hand. "It means, Professor Winters, that I'm dropping this class." He padded to the door without looking back and was gone from our class forever. Perhaps feeling that some explanation was due Fike and me, Winters told us the young chap was a recent graduate of the University of Pennsylvania, all honors student with a brilliant record, but his intelligence had been wasted there for no one on the staff of the English department of the University of Pennsylvania had even the vaguest idea what a poem was.

Winters went on to explain that we would meet here once a week on Tuesday afternoons and each of us would bring at least one poem to class. In the meantime we were expected to attend his course in the English lyric poem which met two early afternoons a week, though we weren't obliged to register for it. If we wanted the credits, he would certainly accept us as students. There was no text for this class save our own poems, but several anthologies of poetry were required for the other course, and we would be expected to read the relevant chapters of his own collections of criticism, *In Defense of Reason* and *The Function of Criticism*. We were dismissed.

Out in the quad I learned that Fike had taken the incident with the young poet from Philadelphia more to heart than I. For one thing they were about the same age, for another he felt Winters' attitude toward him was similar. The old man had told him that he'd chosen him for the fellowship only because "all the other applicants were worse." (I could only hope I wasn't one of the "worse" poets.) When I asked him what he'd thought of the Philadelphia poet, he smiled and said, "Gutsy guy."

Winters' class in the lyric poem was something else. It was conducted in a large airy room whose windows opened on the quad. The teacher sat at the darker end of a long wooden seminar table at which all of the registered students also sat. They numbered fewer than twenty. There were four regular auditors who sat in chairs against the wall; it was a rare meeting that did not bring one or two visitors. Winters always dressed formally. His tastes leaned toward gunmetal gray suits, white shirts, and muted ties, the sort of outfit one expected on an Allstate claims adjuster circa 1957. The graduate students, who made up about half the class, also tended toward formality in dress; many were teaching fellows, and this was Stanford.

The assumptions he made concerning who was in attendance were curious. For example, during the first meeting of the class he remarked that the students would be disappointed if they expected to hear the sort of truths they'd gotten from reading the criticism of T.S. Eliot and John Crowe Ransom. All the students knew that Eliot was the most celebrated poet of the era, but I would guess that none of the graduates had read his criticism. Ransom was better known then than now, but he was hardly a name to conjure with. From Ransom he segued into a diatribe against the prose of Allen Tate and R.P. Blackmur, which he declared was unintelligible. He slammed his fist down on the heavy wooden table and announced that criticism was not meant to replace poetry or prose fiction; its functions were to elucidate and evaluate, and the second function was the first in importance. No one in the class had said otherwise. "I know what you're thinking," he said, "you've come from classes in which these men are regarded as the great minds of the age, and what you're hearing now is heresy." My guess was the students were wondering what the hell they were in doing in his course.

He then passed out mimeographed sheets containing three poems: a sonnet by Shakespeare, a Renaissance poem entitled "Fine Knacks for Ladies," and the Googe poem "On Money." He asked the students to rate them, but only to themselves. He then read them aloud in his monumental style. Whenever he read poetry, he pitched his voice at its lowest and chanted in a monotone, always coming to a heavy pause at the end of the line. His reading style was meant to underline the differences between speech and poetry, and nothing about it was meant to entertain. Poetry was a great source of moral suasion; forget that and you missed the whole point. It's possible the class did miss the whole point, for his voice and reading style so dominated the language itself that the poems sounded much alike. To my surprise he preferred the sonnet and liked least the Googe poem. "Fine Knacks for Ladies" he found delicate and lovely, a fine poem of its kind. He declared Shakespeare's rhythmic mastery far beyond the other two poets', and the poem was serious. I would later discover he could hear poetry as acutely as anyone I would ever encounter, though he did not always prefer what sounded best. How else explain his preference for Fulke Greville to Ben Jonson? When the class ended I had an inkling of how much I could learn from this man.

Encouragement came from an unexpected place. I had sent my friend Henri Coulette two of my syllabic poems, "Small Game" and "Night Thoughts over a Sick Child," and Coulette wrote back from Iowa City that not only he but several other poets as well were excited by them. Even Don Justice was fascinated by their movement. They were probably the first original poems I'd written. When I'd switched from traditional meters to syllabics, something seemed to have been released, and without any preamble I was writing in my own voice. Coulette was anxious to learn from me how I'd managed to handle this form so quickly and with so much confidence. The truth was I had no idea. I had painstakingly copied all of the syllabic poems of Elizabeth Daryush, and then I made some tentative stabs at a few poems of my own in the form, and one day I was writing with this new authority.

"Small Game" was the first one I showed to Winters. He studied it at great length in silence. At last he looked up and said, "The rhymes are very good. The syllabic movement is fine, very fine. A poem in syllabics is seldom very good, but the details here are wonderfully observed." He paused for a puff at the unlighted pipe and said, "What is it about?"

I was caught off guard and, looking down at my own copy of the poem, improvised. "It's about the man who speaks in the poem and the life he leads and the relationship of that odd life to whoever might read the poem."

Winters humphed. "You could probably say that about dozens of poems. I'm not sure you've said anything." I thought the poem was so clear it needed no explanation. I didn't dare quote the MacLeish line about poems simply being.

Before Winters handed the poem back, he asked me to repeat my explanation, and I fumbled through it again. He nodded. When the quarter ended, he sent me a one-page report on my progress. Again he stated that the rhymes and movement were very good, the details well observed, but he could no longer figure out what the poem was about. He recalled that when I'd explained it he had half-understood me, but he no longer recalled what I'd said. He added that I might better use my new-gained skill with syllabics to write poems that were about something. I am still unsure what

Winters meant by "about something." The poem was about something the way poems are about something, though it did not employ the language of abstract thought that was so dear to him. It would be too easy to say he despised the particulars of our lives and thus the language that presented them. I knew that was untrue. I had seen the man in his garden and walked with him through the dappled light that fell from his trees too many times to doubt his pride in what grew and his love for, growing things. He did not feel awkward and unlovable when he bowed to his strawberries and his tomatoes; he did not feel in any way threatened when he yanked tenderly on a branch of his favorite olive tree and spoke of the pleasure its fruit, properly cured, had given him. Calm and peaceful as it was on an afternoon in Los Altos, this was not Eden Garden; it was in the here and now, and Winters fought the grower's common pests—the snails, the aphids, and tomato bugs—and mostly he won. Perhaps "the mid-day air was swarming/With the metaphysical changes that occur," but the dust that rose was that of common earth worked by a man. At their best his own poems testify to what he often did not, that "the greatest poverty is not to live/In a physical world."

Even then I understood his distrust of poems in syllabics. The writing often comes in a great rush, it finds the rhymes quickly, and the poems take their own course. Winters had written a great deal of free verse, some of it gorgeous, but he had disowned the best of it and in some cases even reworked the same material in ponderous couplets. He believed in the morality of form, in the struggle of reason to discover what the imagination had gone in search of. In syllabics and even more in free verse the intuitions seize the poem and direct it, and Winters was frankly distrustful if not fearful of the intuitions. More than once he'd insisted that it was through his abiding trust in his intuitions that Hart Crane had come to his sad and watery end. It could not, he repeated, have ended any other way. According to Winters, all who wrote poetry flirted with madness and self-destruction: the more powerful the imagination the greater the danger. To survive one practiced a heroic vigilance. All the days I knew him he lived that vigilance.

That year Winters gave his first public reading in many years. It was held in the auditorium of the San Francisco Museum of Modern Art, and the proceeds went to the NAACP. (Much to the surprise of many of his readers and in sharp contrast to his clones today, Winters was a liberal as well as an ardent believer in equality, racial and otherwise.) A few nights before the reading he phoned and asked if I would help with his preparation. He asked me to make a list of those poems I thought essential to a final presentation. My list was modest: "To the Holy Spirit," "At the San Francisco Airport," "On a View of Pasadena from the Hills," "Sir Gawaine and the Green Knight," and "By the Road to the Airbase." I'd written the names out on a piece of paper which I handed to him. "Is that the whole list?" he said, "I'm not going to be able to fill an hour with these."

"Of course not," I said, "This is Just a core of what I believe you must read." I realized I'd goofed again. "I'd like to hear you read the Theseus poems and Marie Louise's favorite, 'Manzanita.'" And then I recalled "The Journey," much of which I liked enormously. I suggested that for the sake of variety he might read the free verse version.

"What free verse version?" he asked, looking at me in alarm.

I explained that in an anthology in the Stanford library, a book with a title like *The American Caravan*, I'd found an early free verse sequence which included passages that he'd reworked into "The Journey." He slapped his forehead. "You found that in our library. How did you locate it?" There'd been no trick to locating it, for it was listed in the card catalogue under his name. "My God," he said, "I thought all that was dead and buried. I wrote that over thirty years ago. It was the best I could do at the time. I shouldn't still be hounded by it." I said truthfully that I thought it was wonderful free verse, though I did not add that it was clearly written under the influence of Williams. "Yes, I knew what I was doing, which is more than I can say for most poets writing free verse."

When Winters went off to fix himself a drink, I looked at the little *Collected Poems* published in the early fifties by Alan Swallow. Fewer than 150 pages with notes, it is an ugly duckling of a book. It seemed perfect for Winters. All the words were there and in the proper order, and the binding is sturdy enough to keep things together for a lifetime. When he returned, Winters asked me to listen to a reading of the Theseus poems, a long, brilliant sequence full of gore and sex which traces the life of the Athenian hero from his youthful slaughter of virgins to his exile in old age and his final betrayal. Winters read mainly in his familiar monotone, though now and then the power of the blank verse would seize him and his voice would ride out on a dazzling riff. As he read I asked myself, Who is he talking about if not himself. I knew the poem was written before he was 40, and yet it read like a preparation for the end, the finale. What *chutzpah*, I thought. He'd even outdone Tennyson with his aged Ulysses. When he'd finished, he sat in silence for some time. "No," he said, "I can't read it."

"You've got to read it. It sounds marvellous." It had, and I couldn't believe he didn't know it. "It's superb blank verse and like nothing else." I wasn't flattering him, and he knew it.

"I couldn't read it before all those people." He went through "To the Holy Spirit," which he read in so quiet a voice that I had to bend to listen. Finally he read the little poem that closes the book, "To the Moon," whom he addresses as the "Goddess of poetry." The poem, written before he was 50, ends,

> What brings me here? Old age.
> Here is the written page.
> What is your pleasure now?

Much to the shock of his students, even in class he could not resist discussing his impending death. "That's all," he said, "I'm losing my voice." Then before I could rise to leave, he added, "It's not much, is it?"

"I thought you read very well," I said.

"I don't mean that. I'll read well enough. I mean this little book of poems." He held it in his open palm and shook his head. "I don't think it's enough, I don't think it will last. The criticism is a solid achievement, but this is too slight. I can barely fill an hour with poetry. What do you think?" He was asking me!

"They are true poems," I said, "they will last." He shrugged and looked unconvinced. □

Philip Levine taught for many years at Fresno State and now divides his time between Fresno and Brooklyn. His most recent collection is Breath *(Knopf).*

SEX & DEATH AT UDUB

Kris Saknussemm

*I*t's a bleak gray Boeing day in 1982. A slow mist falls like reminder notices for overdue books, and there it is, in the mailroom of Padelford Hall, home of the University of Washington's English department Scotch-taped to the pigeonholes, in simple black felt pen: DOES ANYONE HAVE AN ELECTRIC RAZOR FOR RICHARD HUGO?

It shocks me, because the Richard Hugo in question is a famous American poet, judge of the Yale Younger Poets series and author of a whole shelf of neat books. I leave without checking my mail.

Hugo is back home in Seattle, having lived in Montana and Iowa and the Isle of Skye, and he is dying of leukemia. From his hospital bed, perhaps he can look out over the battlements of I-5 and see the ferries to Bainbridge and Vashon Islands, the red cranes and derricks in the shipping yards—or, on the other side of the city, traffic strung out over the Evergreen Point Floating Bridge? I don't even know what hospital he's in. Harborview Medical Center? Does it matter? He just needs a shave.

He's a local boy from White Center/West Seattle—a former student of Theodore Roethke at the University, like James Wright. (I remember feeling so proud that my first poems published in *The Hudson Review* appeared alongside some of Wright's last.) Roethke's reputation still lingers around the campus, even though morale in the English department is appalling. The glory days are gone.

That Richard Hugo would be dying does not surprise me. What bothers me is that pathetic plea for an electric razor. Wouldn't a man, a writer of his standing, have an electric razor? Couldn't someone buy one for him? Should I buy one for him? Should I go right down to Nordstrom's and buy a brand new Norelco electric for Richard Hugo? Shiny stainless steel with three floating heads?

What about his family? His friends? And then I think that we in the English department are his friends, his family. But how can that be? He has never given a reading in my time there. Never visited a class or wandered the halls, that I know of. Perhaps he's a prick, leukemia or not. A boozy technical writer of a trout fisherman who learned how to con Eastern academics with images of derelict mining-town taverns full of ruptured old-timers and abandoned women who smell of dirty diapers.

And how am I in the English department, anyway? I'm a graduate student teaching rhetoric to pay my way. Why should I feel guilty about Richard Hugo? Just because I was given his best book, *What Thou Lovest Well Remains American,* for one of the awards I won back in college—and read each word as slowly as I'd sip hot soup?

Richard Hugo died a short time after I started fooling around with one of my students, an African goddess—not in some net-shrouded four-poster bed by the light of a hurricane lamp, but in the front seat of my then-girlfriend's-and-later-wife's VW Rabbit in the huge anonymous parking lot below the campus, often with the rain thundering down on us like lug nuts and bicycle chains, breath clouding the windows. (Once, her big ass pushed one of my Spanish language cassettes into the tape recorder when I had the ignition on to run the heater, and, while she straddled me, we kept hearing in perfect rhythm, "Por favor! Por favor!").

Now I'm sad about Dick Hugo. I think of lines like, "When you leave here, leave in a flashy car and wave goodbye. You are a stranger every day." Why couldn't the hospital staff shave him? When I was an orderly, I shaved men. True, it was usually their pubic hair in preparation for surgery, but the point is there were plenty of electric razors around. "You can prune the shrubbery, but leave the standing timber," one man tried to joke, a tall glum high-school principal with a flaccid bratwurst lying sullenly between his legs.

I'm fairly certain Richard Hugo didn't have his crotch shaved. It wouldn't have done any good for his condition. And it wouldn't have been fair, seeing that no one would shave his face. A famous poet. That was the thing that got me. That's what has taken me 19 years to process. Sure, every once in a while there's a presidential inauguration and they trot out some vintage champion of the officially sanctioned word—but outside these dismal ceremonial occasions and the flea-market realm of the popular song, poetry has become a sanitized, domesticated, museumized, dwindling folk art. Not the language of magic, the genesis code of all human striving, but a neutered cliché.

I could not accept that then. I believed that Walt Whitman spoke to me personally through a battered Sony Walkman. I believed God or the Devil might very well be a brunette in tight jeans and Florentine leather boots, who squealed in foul-mouthed conniption when I did her from behind. She had a husky naughty-sorority-sister laugh and absolutely no idea of what to do with herself; I can smell her pussy and her avocado shampoo even now.

Back then, I believed that everyone, from the Iranian waiter at Broadway Joe's to the Laotian gardeners beavering around the mansions of Lake Washington, was as impressed as I was by Dylan Thomas's grave in Westminster Abbey, with the concluding lines from "Fern Hill" laid into the floor: "Time held me green and dying, though I sang in my chains like the sea."

I'm not saying we should've given Richard Hugo a parade. I don't even know how good a poet he really was, how important he is now that poetry is about as culturally significant as pottery. The truth is I can't think about him at all without thinking about those lost days of my own in Seattle. The city where Thomas Wolfe died. Where Jimi Hendrix and Bruce Lee are buried.

I see the windows of the Swedish Hospital shining in the late afternoon light. A seaplane landing on Lake Union. Skid Row ghosts of old loggers and longshoremen wandering somnolently across Pioneer Square. And me going bowling every morning in the basement of The Hub, the Student Union building, before teaching my composition class. "Zen bowling" I called it, as I wrapped the blindfold around my head, swigging espresso and giggling like a maniac.

Once, I taught a class in the same room where Roethke had held court—the same room where Richard Hugo and James Wright had sat taking notes. I was so excited my first day, I got to class early and then had to sit out on a bench, smoking. I said to myself, "This is an important day. This day will set the tone. I will see visions and signs."

All the other courses I taught were in the Mechanical Engineering Building. Why English would be taught in the Mechanical Engineering Building I have no idea. It was just another of those surreal disjunctions that I perversely savored. To highlight it, I made sure that if there were any equations on the blackboard I always left them there, writing my notes and bullet points around them. It amused me that the students would have to confront this collage of mathematical symbols and literary definitions.

I don't know what my students made of this—and I didn't care. What I was interested in were the hot girls, of which I had many. It was all I could do to keep the door open during office hours. Many of the babes were literally bursting out of their blouses—glossy lipsticked, big-haired, frat-row bombshells and arty goths with garage-sale hats and black widow's gloves—all with asses as smooth and curving as Edward Weston's bell peppers.

I had Chinese, Vietnamese, Cambodians, Nicaraguans, Samoans, Alaskans, and a whole lot of round fleshy white girls with straight hair. These latter were mostly locals. They came from Enumclaw and Puyallup, Wenatchee, Toppenish, Cle Elum, and Snoqualmie. They wore cottontail underwear, chewed sugarless gum, and took showers twice a day. They did not care much about English. They cared about not throwing up at parties. They cared about their weight and their wardrobe. They cared about not getting pregnant and one day finding a good job. Many I suspect had toy animals in their rooms. Most stared surreptitiously at my crotch when I lectured, and all could be made to blush and giggle like flicking on a switch. (The black girls were different. They were louder, more adamant—and blatantly horny.)

I had a hook-nosed Persian princess with a 40D cup giving me flowers and books, and a freckle-breasted hourglass redhead from Moses Lake writing me the most ungrammatical and touching love letters. I'd sometimes masturbate over them and burst into laughter and tears all at once. It was terribly flattering and exhilarating, and my manic passion for them drove me deep into my peers in the graduate program. Torrie, a raving feminist nymphomaniac with trick legs and a mood disorder. Jacqueline, who wrote sad/funny stories she would read to me while sharing souvlaki and beer in Gas Works Park—forever debating whether she should go back to her boyfriend in Olympia—and then banging me breathless in her studio apartment in Wallingford filled with books by Willa Cather and Erica Jong.

I had a one-night stand with a witch who lived in Underground Seattle. She lived on salmon heads and fresh testicles she'd wok-fry in scalding sesame oil. And who could forget sweet Jane? Gifted with the firmest, shapeliest breasts I have ever fondled, but a complete and utter loon, who would later try to commit suicide just as I was coming over—and so I had to break down the front door of her house in Ravenna and drive her to the University Hospital emergency room to get her stomach pumped.

There were countless other little frolics. Barely remembered gropes and pokes at parties. Pulling the wings off nurses. Holding hands in Queen Anne. Anal sex in Ballard (there's a title for you). Midnight drives down Aurora Avenue—the haunted cocktail lounges and repossessed appliances, gun shops, car lots, and hotbox motels. Who can remember it all? Blowjobs. Bergman films. Big plans. I sank dick into pussy like burying fish to grow corn, as my old friend Trey Boyer would say. I could ejaculate luminous jellyfish in those days. Milky-wet sturgeon thrashing on the floor. I was so distraught at the loss of love from the woman I lived with, I did anything I could to get back at her—but even at my most insane, I could never keep pace with her betrayals.

The real-estate agent who was handling the mansion the demon girlfriend and I were caretaking once caught me sunbathing naked on the sprawling back lawn, reading a book. She was the bitchiest agent of them all, too—a flabby old Cadillac-driving prom queen turned lush, and there I was lying on the grass with a hard-on. She looked down on it with an expression of pure disgust, then she hiked up her dress, whipped off her panties, and sat down on it like the world was coming to an end. Not one intelligible word. A 55-year-old woman clawing me so I had to wear a heavy T-shirt to bed for almost two weeks. She got so wet it was like trying to paddle a kayak up a waterfall. I never saw her again, except in dreams.

Just as I never saw the Norwegian plasterer again. I had to call him in after a pipe burst in one of the guest bathrooms. He would've been about 35, but he had a Nordic babyface that made him look much younger. I was sunbaking out on the roof, reading *The Odyssey,* covered in Johnson's Baby Oil, when he appeared in the window. I hadn't even heard his truck pull in. I don't know how it happened. How does anything like that happen? He had a thick doughy cock that became as hard as a marble rolling pin. He went slowly at first, like a big boy in wooden shoes learning to waltz, but at the end I thought we both might go tumbling off the roof into the rose garden, his giant slab hands on my shoulders, me crying out when I came, a hundred white sailboats flecked across the blue lake beyond.

The shame and strangeness. The uncomfortable satisfaction. I took a long hot bath after, and drank half a bottle of brandy. That night, I fucked my girlfriend five times, ramming and reaming her until I thought the intensity of her orgasms might rupture her very being—a continuous explosion of ectoplasm and honey that left her slippery with sweat, pliant and twitching. "Christ!" she wheezed in the dark of early morning, "What got into you?"

Today, no one knows or cares what a good forklift driver I once was, or what a lousy janitor, or that I worked in the railyards in Massachusetts, ripping through a pair of workgloves a day, slugging down the glassiest sledgehammer homemade vodka you've ever not tasted from a Bionic Woman plastic thermos with a group of mad Russians who could fix absolutely anything—except, I suspected, their own questionable visa status—all of us ranting about Dostoyevsky, chess, and Kentucky Fried Chicken. Or huddling around drums of coalfires in the roundtable sheds where these demented young Irishmen from Revere would fight barefisted, sweating like horses, while money changed hands and their blood spattered on the concrete. And, after the fight, you could go out and break off an icicle and hit fungoes with snowballs into the dark. And the smell of steamed cabbage and diesel would be thick in the air.

Such innocence then. Such inspired depravity. Roasting whole giraffes and setting off fireworks. Big painterly fits of pasta and red wine. Pubic readings. All sorts of black-silk-stocking laudanum seances with Byron by candlelight and Anne Rice dripping off the wall. But mainly they were days of hope and longing, of idealism and unexpected intimacies. To lie in bed after sex and discuss Roland Barthes. To sit on the rail of the Ferry Terminal with a newspaper of boiled prawns and thick wedges of lemon listening to a 34-year-old woman with the breasts of an 18-year-old tell how she accidentally set fire to her house when she was six—and how her father chased her down the street with a hatchet when he found out.

I listened to many confessions and private theories in those days and nights. And I read great books. I read *Moby-Dick* in my canoe, or lying on the boat dock on Lake Washington. The night I finished the book by flashlight, a full moon rose and I went up to the house and opened all the windows and put on a tape of whale songs and turned up the stereo as high as it would go: to hear those immense mammals sing with the moonlight flooding in. (Until our neighbors called the cops.)

I met Raymond Carver when he came to read. He was famous by then. He was not a good reader, but he was very funny afterwards when we went to a professor's house for a party. He was the only one who didn't drink.

I had a long and involved conversation with poet laureate-to-be Mark Strand over warm white wine. He complained about the absence of acceptable prosciutto in Salt Lake City. And, with a straight face, he proclaimed himself to be the city's most significant cultural attraction.

And I smoked marijuana with Stanley Elkin in his motel room, both of us breaking out in spasms of weeping laughter. He'd come out from St. Louis for a writers' conference and I was his driver and assistant. "I'm old enough to be your mother," he quipped when, for some reason, I mentioned the year she was born. He got high to cope with the MS he was crippled with. He told me about a fight he got into at a diner outside Chicago when he was just a squirt. Some guy called him "Ikey," and Stanley threw him across a table. He said it was the best moment in his life.

I never met Richard Hugo, who may have died unshaven, for all I know. "What endures is what we have neglected," he wrote, although I feel as if I heard him say it in a bar—to me. All these years later, I'm still wondering what it means, and what became of those lost colleagues and lovers of mine. □

Kris Saknussemm lives in the Pacific Northwest and Australia. His first novel, Zanesville, *the first in a trilogy, will be published by Villard in September. E-mail: krisjohn@vic.chariot.net.au*

CARLA'S STORY

Peter Coyote

I met Carla in 1968 when she was 17, a big, voluptuous teenager with a throaty laugh and a baby. I was 27, the de facto headman of the Free Family commune in Olema. I lost track of her around 1971, when our truck caravan broke up in Boulder. My father had just died and I went East to help my mother. Others scattered to their own needs and purposes, Carla with them.

I ran into her once by chance in San Rafael around 1975. We went back to her room to share a bag of dope and catch up. After that, I lost track of her for 16 years until she called me one day, out of the blue. We met and talked for hours, breathless with the good fortune of having found one another again. This is largely her story, but it is also mine.

It was autumn. I had moved my lady, Sam, and our daughter, Ariel, out of the overcrowded Olema ranch house into a small, abandoned outbuilding, a cattle shed we had tarpapered against the winter, insulating the windows with plastic and the rough wood floors with old carpets. I had installed a wood-burning stove and built a loft for sleeping. It was lovely lying in bed at night, listening to the dull comforting murmur of the winter rains puttering on the slate roof.

According to Carla, I was away at Black Bear Ranch, another Free Family commune deep in the Trinity-Siskiyou Wilderness, when she and her gangly, boisterous boyfriend, Jeff, arrived. Jeff had already been living at Olema and had been introduced to Carla through a mutual friend from one of Carla's foster families. Carla gleaned from others that I was a "significant" person and that she would need my okay to stay. I was the nominal headman by virtue of having been the first to colonize the place and supplying the overarching vision of Olema's dovetail with the rest of the Free Family. It was commonly agreed by all who lived there that Olema was "free turf"; one could do and be there whatever one chose.

Carla remembers that I made her feel at home when I returned and that Sam had appeared to her as omnicompetent and everything she might ever want to be. Sam was ten years older than Carla, a tall blonde from Shreveport whose family had once groomed her for beauty pageants. She was beautiful and imposing, with witchy powers. Emmett Grogan used to call her "the swamp bitch." Like most of us, she was just picking her way through the rubble of her own psyche; to Carla, she appeared as a goddess, fully formed and worthy of emulation.

Sam taught her to tan the deer hides we retrieved in numbers from the Pt. Reyes garbage dump during hunting season; to make an oatmeal-thick mash from wood ashes and water to slip the hair; to pickle the skins in a sulfuric acid bath or rub them with brains, and break them to softness over a fence post or the back of an ax driven into a stump. Though the ultimate utility of such skills might have been marginal, they contributed to a sense of independence from the larger culture and supported our intentions to be in continuity with indigenous people who had lived there centuries earlier. Such skills also enabled us to create tradegoods and currency from found objects, personal effort and time. One could create wealth by redefining it in a game that was not stacked against you.

Carla's deficiencies as an immature mother were absorbed by Phyllis and other members of the community who would grab her baby, Malachi, and take him off with them on their errands and whims for hours at a time. Carla was stupefied and relieved at this display of group concern and generosity, not surprisingly, considering her own memories of home. Her parents were both teachers. Her mother was also a fairly talented painter of portraits, the type that grace middle-class homes, implying status and discretionary income. Her stepfather was discovered with pornographic photos of some of the girls in his class. Carla remembers him as a "sick son of a bitch." One day when Carla was 14, he ripped off her blouse and pinned her to the living room floor. Her mother walked in, regarded them for an instant, then continued to her own room, slamming the door without a word. That was when Carla began running away.

Before long Carla was stitched seamlessly to the rest of us. She participated in co-creating our group mythology and she inspired others, once serving as the inspiration for one of Lew Welch's best throwaway lines. Lew was a famous Beat poet, a tall, freckled, sad-eyed Irishman whose characteristic expression was childish wonder and delight. He lived sporadically and stormily with a thick, powerful Slavic woman named Magda. When she tired of his drunken escapades, he would move out of their Marin City pad and in with us. He loved jazz and Magda's two children and was proud of tutoring their musical abilities. He introduced me to Magda's oldest, Huey, when he was ten years old. Lew beamed with delight as the child scatted the tricky jazz riffs Lew had taught him. He might well have been proud: Lew's Huey grew up to be his own Huey Lewis and the News.

I was pleased to have Lew at the ranch, because I felt his presence conferred on us a legitimate descent from the Beats. I was also extremely proud that Lew had dedicated a poem to me, "Olema Satori."

On the day he honored Carla, Lew was sitting on the floor of the Olema living room cradling a jug of Cribari red in his lap. He was deep in his cups. The room was pulsing with music and dense with marijuana smoke. Carla was lost in sinuous dancing, naked from the waist up. Lew watched her with undisguised lust. He turned to me, grinning crookedly, raised one finger, and said, very slowly and very clearly, "The...worst...Persian...voluptuary...could ...not...imagine...our...most...ordinary...day." Having managed this, he pitched over, unconscious.

There was no clue in Lew's joy that day, that not long afterwards he would leave his wallet and a note in Gary Snyder's kitchen and walk into the Sierra foothills with his rifle, to commit suicide. If he had hidden his private griefs in life, he remained consistent in death. To this day, his body has never been found.

About this time, Rick "Doc" Holiday, a small, delicate junkie with black hair, a con man's politesse, and a soft, smokey voice, showed up with a tall, androgynously beautiful girl who might have been Mick Jagger's sister. This was Daney. Junk-sick as she was when she arrived, she radiated sexual energy and Carla took to her immediately.

Basically, Doc abandoned her with us to clean up. Daney did her best, and before long she was baking bread, laughing, throwing sparks off her cat's-eyes, and radiating a feral energy that made it apparent that she would not linger very long in the land of brown rice and black beans. Both girls were ready for a break from the poverty of ranch life when Carla came and asked me for some money to get high.

The reason she asked me was because, that day, I was in charge of the Free Bank, a Digger institution: everyone put their money, food stamps, and personal wealth into a kitty, to be divided up at group meetings according to consensual priorities. What was left after the kids, trucks, and groceries had been covered, could be taken out on an as-needed basis, and simply recorded in the Free Bankbook.

I tried to talk Carla out of doing drugs, but she and Daney were already gassed up and idling at half-throttle. They left together, Carla in her work boots, floor-length skirt, and flannel shirt, baby on her hip, and Daney, looking fabulous and "reeking of sex," as Carla remembers. Daney did not have any money for herself, and Carla overlooked her assertion that she would "get some" when they arrived in Sausalito.

The two girls and the baby hitchhiked to the Trident Restaurant, a favorite hangout for drug dealers and wannabes. Daney was in her element there, gliding into the room like a shark, leaving Carla at the bar while she went "to score some money."

"I couldn't figure out how," Carla said incredulously. "Even when I saw her crawling out from under a tablecloth, some slickster handing her some bread, I still didn't get it!"

The girls left with an archetypical dope dealer in tight leathers, roaring through town in his BMW-with-sound-system-to-blow-out-windows. In his bachelor pad in the Sausalito hills, Carla filled the sunken tub while Daney went into the bedroom and fucked the guy into unconsciousness. After putting him to sleep, she joined Carla for a luxurious soak while they waited for their heroin to be delivered. While lounging in the suds, Daney confessed her occupation to Carla, who, far from being shocked, was impressed.

"Hey, it didn't sound bad to me at all! Great pads, good cars, easy money, and all the dope you want, being delivered! I couldn't believe it." She lay back in the luxurious hot water, playing with the baby in the shimmering, scented bubble bath, and put that idea on hold for use at a later date.

In April of 1970, we were evicted from Olema. A new cowboy had leased the pastures and didn't want to share them with 30 hippies. We left peacefully, cleaning the house and grounds, down to the last cigarette butt and bottle cap, paying every outstanding bill in town. The citizens of Pt. Reyes and the new lessee understood and appreciated the gesture. While we were definitely freaks to them, they liked us. We had been honest in our business dealings and had certainly supplied ample entertainment and gossip. Tom Quinn, the new lessee's brother, was a local commercial artist. He made an elegant wooden sign with a coyote footprint painted on it. Under the footprint he wrote "...and company have gone." As we drove out for the last time, I wrote the word "on" after the "gone." Six months later, I returned and took the sign itself. I still have it.

The Free Family was preparing a caravan at that time. Our idea was to travel to far-flung locales and use our neutrality as newcomers to create meetings, détentes, and political alliances among people who should, but did not, know one another. Things were to commence with a trip to southern Colorado for a peyote meeting with members of the Red Rockers and the Triple A communes in the Huerfano Valley.

I loved the act of preparing my truck for such trips. Each task I accomplished inventoried a useful skill I had developed since leaving college with a degree in English literature. I had discovered a passion for the deductive, problem-solving capacities required to live without the money to hire professionals to solve life's inevitable dilemmas. My engine had been lovingly rebuilt on my kitchen table; each bolt torqued to specifications and locked against vibrations with Loctite. I had balanced the flywheel and clutch pressure-plate together, and it idled like a whisper. Metal strapping from the bed of an abandoned truck had been arched over the sideboards of mine and covered with canvas, so that my '49 Chevy deuce-and-a-half flat-bed resembled a Conestoga wagon. In honor of the scars and lacerations I incurred during its construction, I named it Dr. Knucklefunky.

Nineteen adults and eleven children trundled nine homemade

house-trucks over the Sierras and on through the sage and scrub of Nevada. Outside Provo, Utah, we camped in a broad flat meadow overlooking a reservoir. It was idyllic: groves of aspens offered pleasant shelter from the wind, the grass was thick and long, the weather balmy. Everyone was having a great time. Everyone but Simon. Simon had joined us from Black Bear. He was tall and skinny with red pimples all over his body, so painful that he took to walking around naked. He also had a huge boil on his tongue that made speech almost impossible. I smashed up some Oregon grape root to make a blood-purifying tonic for him; it seemed to help, but he decided he needed vitamin C and went to town to steal some oranges. He was brought back by an apologetic sheriff. He had tried to be nice to us, he complained, had not hassled us during our stay, but now we had embarrassed him. Simon kept blubbering protestations which were unintelligible due to the boil festering in his mouth. We told him to shut up and apologized to the sheriff, promising to keep Simon confined to camp.

One day, as Simon was wandering around naked, a car nosed along the trail and around a clump of aspen. Simon was either too stupefied from poisoned blood or too arrogantly proud to pay attention and cover himself. The driver floored his car and raced away. As we discovered later, the angry man was the owner of the property and his wife had been sitting next to him.

That night, at around 11, the sheriff, apologetic again, woke us up: the Range Riders were coming to arrest us.

We thanked him for the warning and broke camp by headlights. Only a little trampled grass showed we had been resting there almost a week. We drove down the road to a diner-bar joined at the hip to a small gas station. A Saturday night cowboy frolic was in full swing. While we filled our thermoses and gave the kids hot chocolate, the men parlayed and tried to decide where we would camp. From time to time cowboys would wander out from the bar, survey us drunkenly, then disappear back into the whining maelstrom of the dance. I was beginning to get nervous.

We got everyone loaded into the trucks and were lined up ready to go, when I noticed that Peter Berg and his truck, The Albigensian Ambulance Service, were not among us. I began searching for him. Cowboys clutching pool cues were beginning to cluster in doorways, and I could feel serious trouble was brewing.

I spotted Berg's truck at the gas station and ran over. He was nowhere to be seen, so I pounded on the restroom door. It was flung open and I was greeted by a sight that I was certain would be my next to last on earth. Peter was back-lit in the doorway, wearing his brown leather trench coat, still bald from where his head had been shaved by the Nevada police. His eyes, magnified by his rimless glasses, were crazed by stimulants, and in his hand was a large and bloody butcher knife. Behind him, half in and half out of the blood-stained sink, was what I took to be a flayed human baby!

I imagined one of the cowboys peering over my shoulder and I knew, with paralyzing insight, that we would all be lynched, strung up among the winking bar lights, as a warning to others, the way the locals in that area killed coyotes and casually hung them from fences along the highway.

It was a jack rabbit that Peter was skinning. He had run it over on the way out and had not wanted to waste a tasty addition to a stew. I regained enough voice to convey my urgency to him, and we gathered up the remains and fled, leaving the bloodstained washroom for the locals to ponder.

After four months on the road, Carla was "ready for a hot bath and some dope." She and Jeff traded their Chevy for a little red MGB and piled themselves, Malachi, and all their gear into it and drove straight back to Church Street in San Francisco, where they moved in with Little Paula and the Cockettes.

Little Paula was a short, effervescent brunette I had met during my time at the San Francisco Mime Troupe. She was one of those girls who substitutes aggressive personality for physical beauty. She required thick-lensed glasses that made her eyes appear large and manic, independent of the rest of her face. Since leaving the Mime Troupe, Little Paula had become a skilled criminal, working hot credit cards and using extraordinary amounts of drugs. She also owned a gargantuan tomcat who had been trained to crap in the toilet, a feat that kept the house odor-free and was also guaranteed to stun brain-numbed stragglers who stumbled into the bathroom to confront the cat spread-eagling over the toilet bowl. This was the atmosphere that Carla had been seeking, and it was not long before she had a jones going.

Paula's roommates, the Cockettes, were a drag-queen review that favored Shirley Temple crinolines and tutus, unshaved legs and beards. One of their spectaculars featured Hibiscus reprising Jeanette MacDonald numbers while being pushed on a large, flowered swing. They loved to shock straight people by going shopping while sucking on popsicles shaped like penises.

Like most everyone else in the counterculture, the Cockettes were anti-war activists, and they invented a unique brand of draft resistance. They would pull their van up to the Oakland army depot and offer free blow jobs to young men about to take their physicals. Afterwards, they would offer the lads Polaroid pictures of the event as hard evidence of homosexuality to be presented to the draft officials!

Jeff began to hang around with the Hell's Angels. I don't know whether he was an active prospect for admission or just waiting around hoping to be asked. He was doing B&E's (breaking and entering) and fencing stuff to get by, when Carla became pregnant. They moved out to the suburbs of San Anselmo, where they sold dope for a sweet guy named Kelly, who controlled all the Mexican salt-and-pepper heroin in Marin. He was a stand-up guy; no matter how many times he was ripped off, he took it as the dues that came with the territory. His counts were always fair. He would extend credit. He never ground up the nuggets of brown heroin into the material that it was cut with, so it was easy to pick out the active lumps and throw away the lactose. Everyone liked him.

Kelly's old lady, Carol, however, was hated and feared. A sultry, flashy girl with thick blonde hair, she rode roughshod over Kelly's undying affection. Junkies who got into arguments with her would get cut off. She was famous for leaving people dope-sick and waiting while she shopped for clothes. She didn't like men very much. Her mother had been a hooker, and she had eight brothers and sisters, each by a different father.

One day, while Carla and Jeff were selling for Kelly, the DEA raided their house in one of those B-movie blitzkriegs where furniture is upended, and all the spices are dumped in a huge pile in the middle of the rug. Baby Malachi sat in the middle of the floor with his crayons, coloring diligently, while the house was being dismembered around him. Carla was screaming "would you like me to *open* it for you?" but the cops were oblivious to her ironies. They smashed down doors and shredded pillows, having much too much fun to slow down.

One cop, Jerry, a handsome Kirk Douglas look-alike with shoulder-length hair, was obviously embarrassed by the whole procedure. He sat in a chair covering his face with his hands, repeating over and over, "Guys, you can *see* they're not scared of us. There's obviously nothing here." Carla noted and appreciated his mannerliness and demeanor; coincidentally, he figured prominently in her life a few years later.

The raid helped Carla decide that a future of imprisonment was becoming increasingly probable. Because she could not even consider "the possibility of life without dope," she entered the methadone program in San Rafael, and moved in with Nichole, an occasional girlfriend of mine, who had joined us on The Caravan much to Sam's displeasure. Nichole was a great singer. She was dating Stephen Stills at this time. This would have impressed Carla inordinately, if Nichole had not also taken a shine to Jeff. Sexual generosity, as well as great personal charm, were two of Nichole's endearing qualities, however, so no one ever stayed angry with her for long.

Things went fairly well for a while. Willow was born in that house, with Baby Malachi in attendance, holding his little red wagon prepared with a pillow and blanket to take his baby sister for a ride. "Malachi was the adult in our relationship," Carla used to say. "He adored his sister. Told me when she was hungry, when she needed her diapers changed. He took her everywhere, God bless him, because I could barely take myself anywhere, let alone take care of them." Malachi was three.

With a larger family now, Carla moved to a bigger house in San Rafael. Jeff was working there building trucks with fake compartments for Kelly's drug runs to Mexico. Jeff's fascination with the Angels had continued and he had attached himself to Moose, a gregarious rogue with a quick temper and a steel-trap mind. Moose was everyone's uncle. He visited Olema often in his huge white Cadillac. His Harley-Davidson was white, too, with a large red cross on the gas tank. I always assumed this was because Moose never traveled without medicines for aid and comfort, particularly high-quality methedrine.

He enjoyed "kidnapping" me, as he called it, taking me away for "runs" of days at a time. On one occasion we left so rapidly that he had to stop at Angel Larry's to commandeer a pair of boots for me, solid black Chippewas I still wear.

Moose's real name was Lorenzo and he intimated a connection with the Mafia. He had, according to his own mythology, been imprisoned for life-without-possibility-of-parole at 19 for killing three guys with a screwdriver after they'd made the mistake of jumping him outside a waterfront bar. He *was* awesome when provoked to violence and I considered this story as possibly true. Once, as we were leaving on one of our runs, he asked to look at my scarf. When I took it off and gave it to him, he returned it without looking at it. "If I can get *that* from you," he said sternly, "I can get everything you own." He was full of little epiphanies like that.

Jeff thought he could slide by the Angels' prohibition about needles, and with anyone but Moose he might have, but when Moose discovered needle tracks on Jeff's arm, he beat him so badly he broke a baseball bat over his body.

Carla and Jeff were best buddies at this time, but living separately. Jeff claimed that he was being paid by the Angels, but whether or not his status had formally been elevated to "prospect" was unclear. He was making many trips north to Oregon, and would appear on Friday nights to eat, sleep with Carla, play with his kids, and leave her some rent money.

One day, Moose suggested that Jeff prospect in San Rafael in Marin County instead of Oakland, said that it would be closer to home and less of a strain on him. He was put under the charge of a guy named Red, who ran a gas station there.

Soon afterwards, Jeff came to Carla's truly horrified. He couldn't sit still, couldn't concentrate, couldn't focus his attention. He spoke agitatedly, with big gestures, gulping for air. He kept alluding to something, but all he could actually say was "I'm freaked, Carla. Really, I'm freaked."

She could only determine that Jeff had been wheel-man on an errand with Red and had seen something that scared him beyond measure. He told her that he was going to talk to Red in the morning and that he'd get back to her. He didn't.

The next day, Carla's temporary roommate pulled a robbery that went sour and had to leave town, leaving Carla without one-third of her rent. By the second Friday, without Jeff's contribution, she was down two-thirds and nervous, so she went to the garage to see Red. When she asked about Jeff, Red looked at her blankly and said, "Jeff who?" She knew then that he was dead. She called Moose, her only ally in the Angels, and he gave her a song and dance about Jeff just being gone for a few days.

"I knew this was bullshit," she said, "and I freaked." She went to the police and tried to convince them that her husband had been murdered, "but all they saw was some hysterical biker's broad and laughed me off."

They must have laughed all the way to Red's as well, because the next day an unlit Molotov cocktail crashed through Carla's front window with a note attached to it about not going back to the police.

A few days later, Carla left the kids asleep in the house to slip around the corner to the market. When she returned, all her furniture was piled in the yard with the kids sitting on it, her clothes in a big puddle at their feet. Despite the fact that her rent had always been regular and there had never been any trouble, the landlord's only response to her entreaties was, "You're outta here."

Carla took what she could and left. Despite the help of friends who took her in and gave her dope—without a phone she could not service her regular customers and buy her own—she slept on the streets, in Goodwill boxes, for weeks. Finally, two of her friends, Mitchell Brothers porn stars who had "done a geographic" from some trouble, took pity on Carla, and offered to take care of the kids, until Carla could get her scene back together.

Carla was grateful, but unfortunately one of the girls, the one Carla knew best, got hurt in an accident shortly afterwards, so *she* gave the kids to the *other* girl, Barbara, who promptly moved them to faraway Coos Bay, Oregon.

Relieved of the children for the moment, Carla was scuffling determinedly now, trying to grubstake a house and a means of supporting her drug habit. When she was hitchhiking over the Wolf Grade one day, a fat cat in a big Mercedes picked her up and offered her $50 for a blow job. Carla was stunned.

"Fifty fucking dollars," she thought, "now that's something I can do." So she did, and not only loved the money, but the rush. She liked the rush so much that in later years, even after she and her friends had established a brothel and a substantial client list, she confesses that they would sometimes sneak off into the city and "work guys in cars, for the adrenaline."

She rented a sweet little house in San Rafael, and between

selling dope and turning tricks, paid for a cozy nursery with fresh paint and sweet pictures. She stuffed it with toys, books and pictures, clean bedding, and clothes, and prepared to get her kids back.

When she got to Coos Bay and finally located Barbara, she was horrified. They were living in a filthy tepee. Her children had runny noses and chapped lips, and were covered with mud. Barbara had fallen in love with them and was not about to give them back. She had alerted the town to the threat of Carla's arrival and everywhere Carla went she was tailed by hostile people who had been told God-knows-what about her relationship with the kids—maybe the truth.

Carla finally struck a deal with Barbara to let Carla have her own kids for 30 days. If the children didn't want to live with her after that, Barbara could have them back. Carla returned home with Malachi and Willow.

Two days later, Barbara appeared at their San Rafael doorstep and moved in.

Carla knew that she had to hide her habit and her business dealings from Barbara. To meet her various customers and keep the money rolling in, Carla was forced to make a dozen trips to the store, pleading absentmindedness. She would lock herself into the bathroom and take half her normal dose of heroin so that she would not nod out and drop a burning cigarette into her lap. Finally, the wear-and-tear of inventing excuses and juggling schedules got to be too much, so Carla demanded that Barbara leave and come back at the end of the agreed-upon 30 days.

Barbara asked to be allowed to take the kids to visit her own foster mother first. She said she would return with them that night.

It was not until the next morning that Barbara ended Carla's all-night vigil with the news that she had given both children to the police and told them about Carla's prostitution and drug business. Carla became hysterical. Instead of simply going to the police to report her children had been taken by a crazy babysitter, she made the understandable, but stupid, mistake of going to see a lawyer.

Marvin the Con, as I'll call him, had a penchant for young hookers, and in lieu of money (although Carla estimates that she gave him about twenty grand in cash over the years), he was happy to fuck Carla himself and pimp her to his friends. He gave her a lot of lawyerly advice, which, "if I'd of followed, I'd probably have my kids back today," Carla confesses fairly enough, "but he's still a scumbag."

He sent Carla to a social worker named Jane, a kindly, understanding-looking woman that Carla fell in love with and "just trusted! I told her *everything,*" she says. "I came clean: the drugs, the tricks, the selling, everything." Jane went to a judge and had both children made wards of the court.

Carla moved in with Clint, Kelly's muscular delivery man. The next two years became a blur of moves between every cheap hotel, motel, and rooming house in Marin, until November 2, 1975, when Carla picked up the *Examiner* and saw the front-page photo of a large, algae-covered 50-gallon drum, dripping wet and wrapped in chains, which had just been dredged up from beneath the Richmond Bridge. The caption identified it as her husband's coffin. The police had caught Jeff's "friend," Boneyard, on that bridge with a car-trunk full of cocaine. In searching around for something to deal for his freedom, Boneyard turned over Jeff's final resting place.

The next night, one of Carla's customers told her that the Angels were looking for her. Carla figured that if her *customers* knew the Angels wanted her, it would not be long before she crossed paths with the boys themselves. So for the next month, she and Clint slept in a different place every night. It was nerve-wracking never knowing, when she kept an appointment, whether or not she might find an Angel there waiting for her. Clint and Carla drifted that way for months, skimming the nether world of Marin like fallen leaves before the wind.

By this time, Malachi and Willow were in a foster home in San Anselmo, and they were thriving with a wholesome, nurturing couple who had a yard and rabbits and swings. Carla's appearances there were becoming more and more traumatic. On her final visit, Malachi had clung to her leg screaming and begging, "Take me with you, take me with you, Mommy."

Carla recognized that her life was a shambles; she believed, with good reason, that she could possibly be dead very soon, and so reluctantly, she signed adoption papers, delivering her children to these good people. "I cried for five straight days," she remembers. Clint and Kelly held her, rocked her, fed her, kept her stoned, and never left her alone for five minutes.

Meanwhile, Kelly was on his way to jail. Even the brilliant Terry Hallinan, friend of civil liberties, radical causes, and underdogs, a man who had often helped the Diggers without charge, could not help Kelly this time. Kelly didn't help himself much, either. At his trial when the IRS compounded the charges against him by overvaluing the street price of his dope and demanding $45 in unpaid taxes for each $20 bag of dope he'd allegedly sold, Kelly became irate, jumped out of his chair, and yelled, "If you can get that kinda fucking money for it, I'll sell it to *you*!"

Kelly owed Carla and Clint about 80 grand for bail and lawyers they had extended him during the course of his troubles. As recompense, he introduced them to his primary connection in Mexico, so they could take over his business. All he asked—begged for, actually—was that they "take care of Carol." This was a lot to ask because, as I've mentioned earlier, *everyone* hated Carol. Clint and Carla agreed, nonetheless, so Kelly could go off to prison without worrying.

The night before he went to jail, Kelly rented a big sailboat and catered a haute-cuisine candle-lit dinner with fine tableware and sparkling crystal. He bought matching handmade white doeskin outfits for himself and Carol, and some fabulous jewelry for her. He had prepared a magnificent romantic farewell, but Carol never showed up. Kelly spent his last night of freedom lying on the bed of the boat weeping. At dawn, he delivered himself to San Quentin.

After Kelly left, Carol finally appeared. Carla, infuriated at her cruelty, beat her bloody. In tears, Carol recounted her side of the story—the details of life with a man rendered impotent by junk. Carla was certainly sympathetic to the idea of sex as a basic need, and Carol's tale mollified her and Clint just enough to assure her that they would deliver her maintenance-level quantities of dope. She would have to cover her own rent. Even this was difficult for Carol, because she had grown lazy and dull after years of Kelly's largesse. Carla gave her the phone numbers of some tricks and told her to go to work and see if she liked it.

Not surprisingly, Carol turned out to be a phenomenal hustler. "She got more outta those guys than I could," Carla reports. Carol moved in with a girl named Pam and they started doing "doubles." Then Clint got very hot with the police and had to disappear, so Carla moved in with Carol and Pam. They let a snazzy house on Sunset in Mill Valley, with a lovely view of trees and an easy walk to the park.

Life was good. They had plenty of customers and lots of calls for doubles and triples. A cab would come every morning, with orange juice and donuts, and take them to the methadone program, and all their neighbors liked them, despite of guessing what they were up to, because they were sweet girls with sunny dispositions.

In the curious way that opposites often attract, Carla and Carol fell in love. Carla explains, "I mean we were bathing together, sleeping together, fucking together; getting high together, what do you expect? Besides, Carol was great with me, helping me through bad depressions about my kids and really looking after me. Bein' real sweet."

Clint was incensed about this, but Carla was adamant: if he wanted her, he would have to take Carol as well. Clint did move in, but things didn't work well. One day when Carla returned with the groceries, she found Clint and Carol loading weapons on either side of the living room. This was too much and Carla walked out.

She relented a few days later, picked up Clint, and the two of them moved in together and started a "whole new deal." Carla got a job in a pizza spot in Tiburon, and they both broke their rigs. Alcohol was still allowed and methadone and she still turned an occasional trick for fun-money, but compared to their past, they were almost civilians.

However Clint began killing a fifth of vodka before noon. By evening he'd be blind drunk and dangerous. He was a big, strong guy and gave Carla several serious beatings he would blot out of his mind by the next morning. Carla would present herself at breakfast, black-eyed and puffy, pointing at her ruined face, saying, "This is you, Clint." Clint would insist that he would never treat her that way; he refused to believe that her condition had anything to do with his behavior.

One day, the cops picked her up hitchhiking. They "suggested" she come to the station for a talk. It is a measure of the loyalty that Carla inspires in people that when her favorite trick saw her enter the police car, he risked following her to the police station, demanding to know the charges against her and what her bail would be. The cops assured Carla (and the trick) that she was not under arrest. They told her that they wanted Clint. They had a warrant for his arrest, for a hand-to-hand sale of two ounces of heroin to an undercover cop. They had him cold. Terry Hallinan couldn't help him because of a conflict with Kelly's case; this made Clint crazy as a cutworm. That night, Carla heard him careening down the hallway toward their flat. She decided she'd been beaten enough for one life and hid behind the door. When he stomped into their apartment, calling for her drunkenly, she flattened him with a lamp and fled to her sister's.

Despite their disagreements, Carla didn't want Clint to face his bit in prison without the prospect of conjugal visits, so she married him. The day after their wedding, Clint's trial began. The star witness for the prosecution was the undercover cop that had bought the dope directly from Clint. Carla recognized him as Jerry, the Kirk Douglas look-alike from the earlier raid on her house. She still liked him. "He was just doing his job," she said. "He caught Clint fair and square, nothing personal."

On the witness stand, Jerry kept alluding to his notes. Clint's lawyer rose and told the court that he had petitioned Jerry countless times for these notes and had never received them. Jerry confessed sheepishly that he had recently moved and that during the move the bottom drawer of his file cabinet had become hopelessly jumbled...he had used papers from that bottom drawer to start the first fire in his new house. Clint's lawyer looked at Clint, then at Carla. The prosecutor looked at Jerry. Everyone looked at the judge, who looked at everyone else before he shrugged helplessly and said, "Case dismissed!"

Carla and Clint figured that God had favored them and that perhaps they owed him the commitment of turning over a new leaf. Both got jobs and kicked heroin (again) by using a lot of pot (and methadone) and by going to bars and drinking themselves into a stupor, "because we thought that's what straight people did!" It all took its toll, however, and finally they split up for good.

Carla met a guy at the methadone clinic: a slim, feminine-looking, stab-your-mother street hustler with waist-length black hair, named Gino. He had approached her very aggressively at first and she hadn't liked him, but then for several weeks he had been courteous and polite. One day she left the clinic with Gino, and they went to a hotel and "fucked for two weeks." After a sexless life of several years with Clint, Carla thought that the God of flesh had finally answered her prayers.

Gino's stated occupation was rock'n'roll drummer, but actually he was a con artist. He wasn't above sticking a gun in your ribs, but what he really liked were stings. He was a master at the pigeon drop, the world's oldest switcheroo hustle, but he prided himself on inventing this con: he and a friend would meet sailors coming into port and offer them "girls, any kind you can imagine." He would put together a party of six to eight guys, while his buddy went around the corner to steal a car. On the way to the hotel, he primed the sailors with lurid descriptions of the particular appetites of each girl in his "trap line." They would arrive at a hotel, and Gino would park "for a second" in a no-parking zone. He and his buddy would collect the girls' fees, towel deposits, bribes for the madame and the police and sometimes extra for particular costumes or fetishes he swore excited the girls. Then he and his buddy would enter the hotel and exit through a rear door. The *pièce de résistance* for Gino was that he had also ruined their liberty by leaving the sailors waiting in a stolen car!

Gino also sold fake drugs. He would let oregano sit in a cookie tin for a couple of weeks until it had lost its scent and then mix it with henna and egg yolks. He would scalp tickets to rock concerts, then sell the dope inside. He also robbed gay drug dealers, using his feminine looks and guile to get in the door. He called that one "playing the sugar," because it was so sweet. Once, Carla remembers, "he stole a fag dealer's dog and held it for ransom."

One night, Carla's friend Steve appeared at her door bloodied and shaken, having been stabbed by two black girls while trying to cop dope in Marin City. When one of these girls showed up at Carla's methadone program the next morning, Carla jumped her and pounded the living hell out of her. This was a serious violation of program rules, and not even her counselor Paula McCoy, our old friend and, in her pre-drug days, the most elegant hostess in the hip scene, could save her from a suspension for 30 days.

With the insouciance of the young and naive, Carla told everyone to fuck themselves; that she was going to quit methadone rather than have to put up with their bullshit, and quit she did. It is a testament to Carla's will that she stayed clean and virtually sleepless for *three months,* while Gino was still taking his maintenance doses. When she finally broke and walked into the clinic, she weighed 90 pounds and was shaking like a leaf. Dr. Charlie took one look at her, waived the obligatory two-week waiting period, and gave her an immediate dose.

A little later, the same hard-luck, stabbed-in-Marin-City Steve showed up bloody and ragged *again!* He had propositioned the wife of a guy named Danny in front of Danny's friend Worm. To save his own honor, Danny had smashed Steve in the face with a glass ashtray. Gino retaliated for his friend and kicked Danny's ass publicly in front of a bar on Fourth Street in San Rafael the next day. Unfortunately for Gino, Danny was an unhinged—actually hinge-less—Vietnam vet. After his beating, he returned to Carla's house with a baseball bat and a focused intention to murder Gino. Carla fled out the back door and warned Gino at work. Consequently, Gino was prepared when he encountered Danny later that afternoon and preemptively stabbed him in the chest. Danny lost a lung, but, to his credit, never turned Gino over.

After the stabbing, Gino was too hot to stay in San Rafael, so he fled to New York, leaving Carla a Greyhound bus ticket. She arranged the transfer of her detox clinic (and enough methadone to travel with), and joined with Gino at his mother's in Connecticut.

Gino straightened up and got a job in a warehouse. Carla got a job tending bar at a Howard Johnson's. She walked five miles there and back every day, so they could save toward a place of their own. One night, God smiled on her again. She found a wallet stripped of I.D. with $1,200 in it, in one of her booths at HoJo's. She stashed it in the back and, two weeks later, when no one had claimed it, used the money to bankroll an apartment for her and Gino.

They began a couple of years of holding down several jobs, scrimping to make ends meet, struggling to fend off boredom and despair; a normal working existence. Then, homey normalcy began to pall, and Carla announced her decision to return to California.

Gino did not want to lose her. His dad had worked for a big company, and his mom, moved by Gino's late-blooming domesticity, forged his credit record and denied his arrests on the company's application...so Gino was hired to work for a West Coast branch. He applied all his street smarts and inventiveness to his new work, and today is thriving as one of the company's top service reps....

Back in California, Carla became fixated on the possibility that if there *were* an earthquake, the seismic jitterbugging might create a condition where they would be cut off from their methadone. The doctors at the program pooh-poohed this fear and assured her that all they had to do was turn up at any hospital and demand their dose. Carla's knowledge of the world predicted a different scenario. "I could just see it," she says, "turning up at the trauma ward among the bodies, the wrecked-up and the fucked-up, two junkies looking for a fix. Imagine how long we'd have waited in the back of *that* bus!"

She and Gino decided that it would be prudent to kick methadone in anticipation of "the big one." They tossed a coin, and Gino won (or lost) the toss and quit first. He began slacking off by a couple of milligrams every couple of weeks until he felt normal at that dosage. He'd "keep the edge off" a while and then diminish it another couple milligrams. He began jogging and getting really fit. Carla maintained them both by selling half her doses and the rest of his, and working.

When her turn came, Carla stayed true to her word and cleaned up, although it took her two full years. She then entered a period of intense isolation. Her mother was dying, and she introduced Carla to *The Aquarian Gospel*. Carla began reading anything spiritual she could find, even Jehovah's Witness pamphlets she found on buses. She had no idea how she would be able to live without dope and felt these books might help. One "adjustment problem" was that Carla's newly awakened body remembered sex. She was now constantly aroused, but all of Gino's surplus energy appeared to be dedicated to regaining his physical fitness through exercise. Soon they cashed it in as a couple.

Carla got a job with Marin Towing, a company that hauled away disabled and illegally parked vehicles with their snazzy yellow-and- chrome towtrucks. She felt comfortable there, because the business reminded her of prostitution, especially the litany of services and prices: $25 to unlock; $50 for a straight tow; $75 with dollies.

"It was legal stealing," she laughs. "The cops back it up, even set the rates. A shop owner sets out a little sign that says if you park here your vehicle gets towed. The sign cites some numbers in the public law books, and the towtruck boys are in business, working on straight commission."

She and the boys used to sit on the hills over Sausalito scanning the parking lots with binoculars, looking for illegally parked cars. "Hell, we busted Kenny Rogers's car, and Todd Rundgren's," Carla recalls. "Todd was so impressed that we towed his car correctly that he hired the towtruck driver as his driver. We worked 17 hours a day and I didn't have *time* to be junk sick." Besides the excitement of the work, an added perk was being surrounded with vigorous young drivers: "I fucked everything in sight," she remembers dreamily.

She stayed with Marin Towing for three years, until she was virtually running the office, augmenting her pay checks by towing race cars on weekends. Finally, the owner couldn't afford to pay her what she needed, and, with regrets and great memories, she was forced to leave her first real oasis in many years.

She applied for a job as a cashier at one of America's great brokerage houses. She loved it. Stockbrokers drank like fish and partied hard. They were as unabashedly materialistic as hookers, played all the angles, and according to Carla, worked their customers "just like Johns." Friday through Sunday, she held her demons at bay by drinking herself into oblivion.

One Friday night, her boss gave her his credit card and told her to reward the girls in the office for a tough week by taking them out to party. Carla piled them into her lovingly restored Pontiac Firebird and took them out for the night, firmly resolved to have just one glass of wine and then go home.

"But I can't have just one," Carla says reflectively. By the time the boss joined them, she was so out of control that he took her car keys and put her in a cab. Carla did not want to wake up in Richmond, 30 minutes away, without her car, so she ordered the driver back. She used a hidden key to start her car and head for home.

She doesn't remember much about the trip except smashing into a Volvo as she headed the wrong way down a one-way street. When she finally recovered perception and memory, she was in the middle of the Chevron oil-refinery complex, having crashed through a set of heavy gates, and wrapping the car around herself like extravagant steel clothing.

The firemen took three hours to extract her from the wreckage. Another hour later, at the police station, her blood alcohol measured .23, over twice the legal limit.

Carla called her boss from jail at about four in the morning and told him to leave her there because she needed a vacation.

After all those years and escapades, she was in jail for abusing a legal drug.

Two years ago Carla joined AA. She's found her children and is working hard to repair what can be salvaged of their tattered relationship. She is still stunning. Her shoulder-length hair is punk short in front. Her adolescent baby fat has been burned away, exposing chiseled cheekbones and a slender, aquiline nose. The only trace of her old life I can detect, besides her street smarts, is the excessively polished way she says "Good evening," when she answers her phone. I inquired about that, suspecting that she might still be using the phone for business. She looked at me for a minute. Her dark eyes were as bright and undiminished at 39 as they were at 17. She took a drag of her cigarette. She smiled. "I always thought it would be low-rent to turn tricks after 30, Coyote, so I stopped." □

Peter Coyote is an actor and writer who lives in Mill Valley. His first paid appearance in print was in ZYZZYVA. This piece, published five years later, won a Pushcart Prize and was integrated into his memoir, Sleeping Where I Fall *(Counterpoint Press, New York).*

PRACTICING

Glenn Kurtz

I am sitting down to practice, as I sit down every morning. I tune my instrument, a guitar made from the wood of a door salvaged from a Spanish church. I file the nails on my right hand, first with a steel nail file, then with very fine sandpaper. Even the tiniest ridges can catch on a string and make its tone raspy. I hold the instrument, settle its weight properly, and adjust my body to the familiar contours. I begin.

At first, I just play chords. The sounds feel bulky, as do my hands. I concentrate on the simplest task, to play all the notes at precisely the same moment, with one thought, one motion. It takes a few minutes; sometimes it takes half an hour. But I cannot proceed without this clarity of sound, motion, and thought.

Slowly, the effort awakens my fingers. Slowly, they warm. As they loosen, I break the chords into arpeggios. The same notes, but now spread out, each with its own place, its own demands.

My attention, too, warms and sharpens; I shape the sounds more carefully. I remember now that music is vibration, a disturbance in the air. I remember that music is a kind of breathing, an exchange of energy and excitement. I remember that music is physical, not just in the production of sounds—the instrumentalist's technique—but as an experience. Making music changes your body. I become aware of myself, and I feel as if I've been wandering aimlessly until now, as if all the time I'm not practicing, I'm a sleepwalker.

I calm myself and concentrate. Give the sounds time, let the instrument vibrate. I have to hear the sounds I want before I make them, and I have to let the sounds be what they are. Then I have to hear the difference between what is in my head and what comes from my fingers. As I listen, my fingers make incalculably fine adjustments of angle, speed, and strength of touch. The sounds from the instrument slowly merge with what I hear in my head.

After 20, I begin to recognize myself. My hands feel like my hands and not the mitts I usually walk around with. I recognize my instrument's tone; this is how I sound, for now. I recognize my body: I feel like a musician again, a classical guitarist. I feel ready to work, ready to play.

When I began playing guitar at age seven, playing came easily. There was music in the house—a few guitars, a piano, a flute, an old steel drum my parents had picked up on a Caribbean cruise. My brother had posters of guitarists on the wall—Jimi Hendrix, Jerry Garcia. My mother kept plaster busts on the piano—Beethoven, Chopin. It was 1969. I taught myself a few chords from the diagrams in a Bob Dylan songbook. Pretty soon I could play "Blowin' in the Wind."

The next year, I stood squirming in front of Kent Sidon, director of the Guitar Workshop in Roslyn, Long Island, not far from our house. At the time, the Guitar Workshop taught mostly folk music, although among the teachers were a few classical guitarists. I remember Kent singing a sea chantey, standing with a guitar on his hip, the neck sticking out in front of him, and a cigarette waving like a conductor's baton from the corner of his mouth. "Wiggle your fingers," he said to me, holding my hands palm up in his. I wiggled my fingers. "We usually don't take students younger than ten." But he took me. By the time I was ten, I was a star at the Guitar Workshop.

No one ever forced me to do my lessons. I loved playing. I got better without noticing. It never occurred to me that I was practicing.

After "Blowin' in the Wind" and "This Land Is Your Land," I learned barre chords. I learned the Travis pick. The Travis pick is a right-hand pattern, a basic finger-picking style for folk guitar. You pluck instead of strumming, playing notes instead of chords.

Once you start plucking, the guitar becomes a serious instrument. From the Travis pick to the beginner's etudes of Fernando Sor is just a matter of degree. It took me a year or two. By the time I was 13, I was playing all of the guitar's major composers—Sor, Tárrega, Villa-Lobos. I didn't feel I was learning to play the instrument anymore. I knew how to play. Instead, I was trying to improve, working to become a musician.

Each day, I sat down to make myself better. I practiced scales and etudes. I practiced ear training and music theory. I studied scores and listened over and over to recordings of music and musicians I admired. I listened to the guitarists Julian Bream, John Williams, and Pepe Romero, and to Leonard Bernstein conducting Beethoven and Mahler, to Alicia de Larrocha playing Mozart piano sonatas, to Glenn Gould playing Bach and Schoenberg. To play like them! For four or five or six or ten hours every day, I pursued my path to artistry.

When you play, and playing comes easily, playing is enough. But as you play more, as you learn more about music, about listening, your imagination outpaces your ability. You conceive an ideal music that shimmers ungraspable in the air, or in the hands of others. This music beyond you, as you are, leads you on, and you struggle to lay hold of it. Playing, you've begun to practice. And practice has made "perfect." Now you'll never play the way you wish you could. You'll never be done practicing.

I begin my right-hand exercises. With the metronome set at 50 beats per minute, I play a C-major chord. My right index and middle fingers play the top two notes, C and E, and my thumb begins on the 5th string, playing C an octave lower. Every other metronome beat, I move my thumb, from the 5th to the 4th to the 3rd and back to the 5th string: C, E, G, C.

One and Two and Three and Four and.

I change chords to G^7 and repeat the same right-hand pattern. My index and middle fingers now play D and F, and my thumb moves B, D, G, B.

One and Two and Three and Four and.

Repeat.

It doesn't really matter what chords I play. Sometimes I'll play A major and E^7. Sometimes I just play open strings and let my left hand hang by my side. The exercise is for my right thumb, to practice moving from string to string.

I listen to the strings vibrating and, in the last instant before each downbeat, I plant my thumb and the two fingers on the strings they are about to play. Pull and release. Lightning fast, exactly together, and exactly on the beat, I hope. I concentrate on striking the strings consistently, so that the movement from note to note is smooth, continuous, a line.

After 40 minutes, my right hand feels warm and capable. I feel each of my fingers distinctly, in a halo of ability. I follow the 120 patterns for the right hand written by the 19th-century guitar virtuoso Mauro Giuliani. I work on 20 each day, so that every week I practice all of them. Then I change the metronome setting and begin again. These patterns at 108 beats per minute, rather than 50, demand the same movements, but feel very different. Each of the variations requires concentration and repetition—perhaps months or years of repetition—before I know it will continue and not repeat.

At my high school graduation, I won the George E. Bryant Creativity Award. As I walked off the stage, a friend in the school orchestra tossed me a beer from the cooler he'd hidden inside a tympani. "Way to go!"

He'd said the same thing a few weeks before, at our Spring Concert. I had conducted the orchestra in Fantasia in C Minor, a piece I composed as an independent study project. I'd also performed with the school jazz band, delivering a screaming solo over the "Theme to *Rocky*." What more could high school offer?

I went to the New England Conservatory of Music in Boston and majored in classical guitar performance. The Conservatory was a hothouse, full of strange and exuberant growth. Everybody had won a high school creativity award. But some violinists and pianists possessed freakish, Paganini- and Liszt-like talent. A trumpet player in my class had synesthetic perfect pitch. Each note shone as color in his mind. He would look at a painting and tell us what key it was in.

My teacher had me return to basic technique. How did my fingers meet the strings? Do the left-hand fingers fall cleanly, directly? Or do they fidget on the string before they settle? Does the right hand play the string by touching it? Slapping it? Scratching it? I taped a copy of Leonardo da Vinci's drawings of the arm and hand to my wall. I wanted to know the arrangement of muscles and tendons responsible for each fingerstroke. I practiced in front of a mirror to see the jostling in my forearm, as the large muscles moved the smaller ones. And I discovered, much to my surprise, that not only my forearm, but my neck and mouth and tongue all set to work to make my fingers go.

The best players had a purity of technique. They'd learned the right way from the beginning, then played for a long, long time, according to a single, clear impulse. My technique was confused. The distance from the Travis pick to Fernando Sor was greater than I had imagined. I had to catch up. Even the best players felt they had to catch up. Every day, in a practice room surrounded by practice rooms, I worked for two hours before classes, then for two hours in the afternoon, and again for three or four hours after doing my assignments. I practiced on family vacations. I practiced instead of going on spring break.

In my second year, I won two regional competitions, though I didn't trust the judges. I played solo at an open-air concert before an audience of 8,000 people. Afterwards, a woman told me the ovations were well deserved. I hadn't been able to hear them from the stage. When she asked how I thought it went, I said, "I played the way I play." But I added silently, "for now." I felt wounded by all that had been left out of my performance. The next day, like every day, I sat down to practice.

But there was something corrosive in my striving, in my intense devotion to what I imagined and my constant deprecation of how I played. After four years at the Conservatory, my ear and my imagination became more educated, but my fingers grew tense, almost resentful. I worked and I worked to improve, and yet every morning I awoke still myself, with the same problems, the same tension, the same flaws. The harder I worked, the more my playing fell apart in my hands.

In my late twenties, despairing of a viable career as a classical guitarist, I stopped. I typed and filed for a year at a publishing house. Then I got a graduate degree in literature. I taught digital media. I worked in business. I played only occasionally. I never practiced.

It took ten years for the tension and disappointment to abate. Now, I've started to practice again. This time, not as a young, aspiring artist, but as a former musician. Almost nothing remains of my earlier skill. I still know how to play, but my fingers don't. All those years of work, the devotion and solitude, are like a lost faith, impossible to recover. But I'm not seeking to restore my old ability. Practicing, and my faith in the ideal of that work, led me to despair. I'm trying not to repeat myself. I began to practice again, because I felt I could do it better this time.

I open the music on my music stand. *The Complete Sonatas and Partitas for Unaccompanied Violin* by J. S. Bach. Though central to the violin's repertoire, these pieces sound best on the guitar. The guitar clarifies their harmony and liberates the counterpoint. I've played through this music many times before. The book is fraying from use. I should buy a new copy, but each page is full of pencil markings—fingerings, phrasings, harmonic analyses. It would take months just to transfer these old notes to myself to a new edition. A book of music is a work diary.

I'm working on Sonata No. 2 in A Minor. I play the first movement very slowly, *grave,* listening to the attack and decay of each note. I want to feel the instrument's entire body rising and falling with the sound. I've been practicing this movement for weeks, trying out different shapes, different expressions. The piece is under my fingers and in my ears. But today, instead of pushing farther, I move on to the next movement, the fugue. Phrase by phrase, I'll put it together until, after a few weeks or a month, it is whole. Then I can write the new fingerings and phrase marks on the score to remind myself of how I want to play it.

In a year, I'll go through the whole book. But by the time I return to the first of the sonatas, it will sound different to me. My fingers will have changed; my ears will have changed. Different phrasings and fingerings will suggest themselves. And I'll have different taste in rhythmic and harmonic emphasis.

Musical artistry may seem divine, but practicing is always mundane. Practice immerses you in your daily self—this body, these moods. You sit down to play; you file your fingernails; you shape your reed or rosin your bow. You play scales and exercises. You struggle with mistakes and flaws. The work is physical, intellectual, psychological. It can be exhilarating and aggravating, fulfilling and terribly lonesome. But it is always just you, the instrument, and the music, here, now. And though you sit down to work every day, it may take years before you know what you've practiced. □

Glenn Kurtz lives in San Francisco and teaches at the California College of the Arts. E-mail: gkurtz@digitopia.com

THE VALLEY

Victor Davis Hanson

I will never leave the San Joaquin Valley. I was born here—my children the sixth continuous generation to live in the same house. I will end here, too. I am afraid I can only love what I now so criticize. But the Valley, it is no longer pretty.

For 25 years, I have tried to plead its cause to outsiders. I have argued that it was beautiful, not an ugly place as most believed, and so it deserved some aesthetic repute. Years ago at Santa Cruz and Stanford, I told professors and fellow students (whose Valley horror stories were comprised of little more than car boil-overs in Bakersfield or gas fill-ups on the 99 in transit to Yosemite) that its nights were unusually warm, its sunsets far better than those on the ocean. Out of politeness, I omitted that it was also a region of praxis, not theory, of men and women who with their labor paid the consequences for the ideas of distant others. No one has ever said that this is a laid-back, complacent, or even contemplative province. No one has ever claimed that Americans came here to sightsee, vacation, garden, ski, hike, or shop. The Valley's attraction was always its bustle of hard-working peers, and the bounty that grew out of their endless and uncomplaining work.

The Valley, I summarized, was irrigated trees and vines—a relatively unknown but veritable food basket—in a desert, framed with snow on the Sierra, the evening sun over the Coast Range, and, in between, some half-million absorbed and autonomous ethnics on 40-acre farms.

In Greece, I pointed out, Boeotia was like the farmland outside Visalia, the Attic hills were not unlike those above Fresno, the citrus-raising hamlets of the Peloponnese were similar to Exeter, Orosi, and Porterville. Does not the Valley share the

Hellenic latitude, its climate similarly Mediterranean? Did not Greeks, Italians, Spaniards, Basques, and Armenians as well, come to Fresno for its southern European seasons?

When I returned from one visit overseas, I even planted a bay tree *(laurus nobilis)*, and it grew here next to the vines every bit as well—no, better—than those I saw in the Pindus Mountains. In general, the Mediterranean triad—grape, olive, and grain—also flourishes here.

And compared to Back East, the heat of our summers is dry, and our Aprils and Octobers are not wintry.

In the sorry record of agrarian history, family farming, sustainable agriculture, communities of yeomen peers, and their accompanying cargo of free speech, consensual government, and natural inquiry unfettered by government or religion, are rare historical occurrences. Like the comet in its centuries-long parabolic orbit, these cultures appear only for a while and do so with a radiant burst—in the Greek city-state, Republican Rome, eighteenth- and nineteenth-century Europe and America, and perhaps now again in rural China.

In contrast, palatial bureaucracies and vast estates are the drab norm, where farming is food-production-as-part-of-capital-exploitation. An urbanized population with a surrounding dependent countryside is the more common paradigm, with its detritus of serfdom, peasantry, a wealthy overclass, and political autocracy. Thus, as historians, we must acknowledge how rare was the beautiful century, 1870–1970, when gravity-fed irrigation in hand-dug ditches from the Sierra first turned a weed-infested desert into an oasis of small tree and vine farms and their quiet satellite communities. We must acknowledge that the Mayas, Aztecs, Egyptians, Hittites, and Gauls were not unlike the polis Greeks!

The elder yeoman now dead came into the Valley—I mean the eastern side of the Valley with the water and near the Sierra— and planted and beautified. Like Lucretius' Roman *agricolae,* they "set shoots in branches and buried fresh cuttings in earth about their fields. They tried to grow first one thing, then another, on their loved lands, and saw wild plants turn tame in the soil with coddling and gentle, coaxing care."

Thirty years ago, you could have agreed with Lucretius that here too there were "lands everywhere marked with beauty, lined and adorned with apple trees; and fruitful orchards wall them about." But not now. Rarely are any "lands marked with beauty," rarer still any "gentle, coaxing care."

In the contemporary Valley we are becoming terribly over-populated. By some time in the next century we will have over 20 million suburbanites between Sacramento and Bakersfield. And the farms that survive that long will be like rural atolls in an unending urban sea. Already the air is at once foggy and smoggy in winter. If in the old days we worried about the hundred or so each year infected with Valley fever, the fungus that got into the lungs and then brains of those newcomers unfamiliar with our dust; now the enemy is the atmosphere itself, not the trace toxic spore that floats up from the dirt. The pollens and particulates raised by agriculture combine with industry's ozone and monoxide to create a uniquely ugly haze that gives our people perpetual allergies and asthma, which no fungicide shot into the cranium can cure.

Nothing in this state's history, neither the original forced annexation from Mexico nor the trek of thousands from Oklahoma during the Dust Bowl, can rival the sheer rapidity and totality of the transformation of the Valley. Aeschylus' Prometheus, Sophocles' choral ode in the *Antigone,* and Euripides' Theseus in his *Suppliants* all acknowledge that man can, through artifice, extract bounty untold. But unlike us, the Greeks at least saw that such success can also make man arrogant and worse.

Reader, remember, irrigated valleys, whose rich loams stretch out of flat plains, where water in the mountains above is but 30 miles away, where neither hurricane nor ice intrude, are *not* common on this earth. Perhaps in Chile or South Africa, and maybe in Western Asia Minor and Southern Europe such oases are found, but they are smaller there and less opulent. The San Joaquin is a natural slip, the gods' mistake of enormous proportions.

Our natural advantages, it turns out, served even better those who would not farm, and who saw this Valley's future as a bustling metropolis of teeming urbanites, not a backwater agrarian refuge. These commercialists, inventive and audacious folk all, were not bothered by our reputation for rusticity and yokelism. They welcomed the absence of symphonies and ballets and museums of the first rank in Fresno and Bakersfield, which had once been our aegis, our salvation as a place for bumpkins that most others avoid.

No, as purveyors of inanimate capital, the clever investors rather liked our repute as the Theban dullards of California. They were, in other words, not snobs.

And unlike us, these anti-agrarians were not stupid; they knew how to acquire what was once ours and expand upon what they took. They knew that few farmers would buy out their failed plants and their soon-to-be slums, to rip up concrete and replant orchards. Successful or otherwise, their work remains forever; ours will not. Even in defeat and bankruptcy, the commercial developer wins the race to replace farmland. Rip up a vineyard, and farming ceases—for all time to come. Abandon an unneeded and overrun apartment complex, and a parking lot, storage yard, or simply weed-infested asphalt can take its place. I have never seen a fruit tree or vine replanted on vacant and unwanted commercial real estate, the soil once more set free from its veneer of asphalt.

The developers like—no, adore—this Valley. First, the climate is dry and hot and thus is as good a place to keep roads, trains, and airways open as it is to dry raisins. You can hammer, fabricate, and nail shingles almost every day of the year, just as you can prune or disk year-round. No one is barricaded from work by the snow. Tornadoes do not destroy the box plant, just as they have not uprooted orchards. Earthquakes are but slight tremors beneath our Valley's cushion of loam, shaking neither studs nor barns. The cold does not put welders into depression or prompt one drink too many, as it has not killed vineyards or ruined peach trees. Our university growth institute ("Fresno Futures") confirms to the potential developer that there are 50 days more here of working weather than at his present abode, that an acre of asphalt and aluminum can sprout more jobs than can ten of trees and vines. And it can—and often with less water.

But weather is not the Valley's most attractive enticement. Things are cheap here—food, fiber, and fuel. We are autarchic and grow every type of comestible, $13 billion worth and more each year. You know that our fruits and vegetables feed the nation. But it is forgotten that Fresno and Tulare Counties are the world's richest cotton and dairy centers. Grain, hay, and rice are produced at higher per-acre ratios than anywhere in the world. And there is still oil near Coalinga and Bakersfield. In other words, doing business near Fresno—eating, burning gas, traveling to and fro—is, for a while longer, inexpensive. We supply our own, and ship the natural and manmade surplus to you.

Oh, one last thing caught the eye of the housing magnate. It was also beautiful here, that century of irrigated grapes and tree-fruit, that patchwork of farming peers who turned 40-acre blocks into large gardens, striving for splendor even at the price of economy. It was, in other words, a good place to have backyards border walnut orchards.

And it is not just suburbanites who have been lured here with promises of peach trees and grapes but a mile from their door. The state has located most of its new prisons here, with an equivalent logic. We have many unemployed workers who guard well; we have flat, empty land that is congenial to walls, wire, and cement; we have many criminals who can make the trip short from court to cell. And we are largely free of the Sierra Club, Environmental Defense Fund, and Nature Conservancy, who still ask questions before concrete is poured, who do not value jobs that accrue from the underbelly of modern urban life.

Each little farming community of this eastern corridor of the Valley now has an industrial park and a Chamber of Commerce recruitment team whose job is to snag more suburban light industry ("expanding the tax base") and retailing ("more service-sector jobs") that will ruin their downtowns and pave over their orchards. Years after their success, the more honest main-street hucksters at the brink of the abyss are known to confess—in moments of religiosity, or morose at their retirement dinners, or perhaps cancer-ridden at their final interviews to the local weekly: "You know, I'm still not all that sure it was such a good thing to put that Kmart here," or "I guess we didn't quite realize that we more or less made a whole new town out there by the interstate, and an uglier one that the one we already had." Then they die.

It is ironic that the only vast expanses of uninterrupted greenery left in central California are the cotton, tomato, and rice fields of the west side, the corporate side of the Valley. There, in *latifundia* of 5-, 10-. 20-, and 30,000-acre blocks, are the company towns that have no middle classes, only the massive sheds empty of their land barons who live as absentees in the gated estates of Fresno. There, the growing salinization and the general absence of anything planted taller than fodder ensure that no one wishes to move there. You can drive along the west side, over salty land that should not be tilled or irrigated and does not drain; where no middling citizen exists; where all the water is imported by the Federal Government from hundreds of miles away; where company towns are but bars and ramshackle houses...and see no development there at all. None. Not a hospital, Little League field, Rotary insignia; not an In-and-Out Burger to be found anywhere among the helotage.

We prefer, you see, not to destroy agribusiness, which is not inviting, but agrarianism that is. The salvation of agribusiness is that its physical infrastructure, its very culture, its entire sociology, is so ugly, so repugnant to the heart and mind, so devoid of life in the flesh, that no developer in his right mind would want to plant his subdivision among its empty and godless miles of cotton, wheat, and alfalfa. The curse of the small fruit farmer is that he creates a world that the Los Angeles refugee rather likes and thus will surely destroy.

The San Joaquin Valley is 300 miles long and about 50 to 70 miles wide. For a while longer, on its eastern side, you can still build entire towns, military bases, and chemical-waste sites without running into too many human or natural obstacles—and with no worries from a stubborn and cranky polis of parochial agrarian yeomen. Fresno—arson capital of the nation, the state's leader in per-capita car thefts—sprawls 25 miles, the teeming sewer of Victor Hugo come alive. Developers offer money to the city council members, a number of whom in both Clovis and Fresno are currently either in prison or under indictment. And thus a master plan is rezoned, and suddenly there are another 800 houses.

Farmers are happy to sell, to get that one final harvest worth more than all they have made in a lifetime of financial setbacks. One of the more insightful and happy recent arrivals remarked to me, "This is like L.A. in the fifties, before we wrecked it."

Business has also learned that we natives, like rustics everywhere, are essentially conservative. We are not exactly mindless, but we would rather screw something up first and try to fix it or forget it afterwards—we do both well—than not try something stupid in the first place. So we rarely say no to housing developments, industrial parks, apartment complexes, or freeways. Nor are there any doctrinaire Republicans here, who occasionally practice a noblesse oblige, and whose rarefied sense of art and culture make them sometimes sensitive to ecological damage or to a brutal development next to stately turn-of-the-century homes. No, we are poorer conservatives, mostly yellow-dog democrats (like myself) , an odd mix of ethnic farmers (like my siblings), the children of Okie dust bowlers (like my wife), and those self-made Mexican-Americans who vote religiously (like my sister-in-law, nephews, and friends).

We have know the minimum wage and so are not the sort of people to tell industry to go away. We are not predisposed to worry about brown air, tainted water, or bulldozed vineyards, when it means a steady diet of Burger King, new trucks on credit, and video rentals. My brother-in-law, the insurance agent, told me, "Sure, it may get ugly and crowded, so what? Who gives a rat's ass, anyway? For the smart guy like me, there is going to be a lot of action and money around here for a long time to come."

A farming acquaintance reaching retirement is the antistrophe of that chorus: "No goddamn government is going to tell me who is and who is not going to buy this land. Why, I can cement the whole damn thing over, if that's what I want. I didn't farm for 50 years to make a free park or refuge for someone else. if you feel there's not enough ground left, give your own goddamn farm to some do-gooders, and let some asshole from the government get a big check making sure nobody uses it."

Because of the thousands who arrived from southeast Asia and the hundreds of thousands who came up from Mexico in the eighties, the question of growth is now redefined in culturally sensitive terms. The industrialist and the corporate farmer, their elected aides-de-camp, and the local blunter wits who draw in cheap labor, bark out that a hundred-mile-long megalopolis is the color-blind American Way. "Let all come who would work" is their unabashed credo. Of questionable cultural sensitivity themselves, and aware of the over-cleverness of their no-growth antagonists, they thrown down the gauntlet: "We don't care what color you are, move here if you want to work hard."

Their uneasy allies in the race industry nod, and the professional spokespeople for minorities proffer a variety of stock responses to any who say we are now too crowded in the San Joaquin — as if third-generation Mexican-Americans like sprawl and pollution. "This is but Alta California, anyway, and now we are taking back what is ours." "You are not really worried about the air and the water, but only that those who breathe and drink it now are not white." "We fought for you in Vietnam and so you owe us the home you took."

Quickly, the timid naturalists, preservers, and ecologists apologize—is it not because of "arrogantly thinking that they should know in time"?—mumbling something about race having nothing to do with too many people in too confined an area. Then, like their doomed brethren of Thucydides' time, the more subtle and keen minds go back to their esoteric charts and graphs, which at the next regional meeting to the resource council prove we of all races are finished anyway.

And last, we are just about in the geographical center of California. Shippers love Fresno, halfway between the coast and the mountains, the median between San Francisco and Los Angeles, Sacramento and Bakersfield. The airport is still uncongested, the freeways yet fluid, the city a "gateway to the Sierra," and all criss-crossed by rail. It is not difficult to get in and out of Fresno. Valley roads are paved and wide, and take you away with ease. In other words, in every direction Saroyan's Old Fresno is now less than four hours away from 30 million customers.

Pindar said "water is best." It is. One can judge a society's moral character by the purity of its subterranean water. Ours is now both tainted and increasingly scarce. The aquifer 30 miles from the Sierra drops off to several hundred feet below the surface, and even near the mountains at its shallowest it has begun to fall a few feet every year. Our answer? Drill deeper wells, of course. I am saving up money for yet another 1,500-gallons-a-minute well to water vines faster than those in town can suck out the poor water beneath our feet. On this farm we have lowered all the pumps 20 feet in the last ten years, pumps whose bowls were set at a steady 45 feet 70 years ago. Our droughts of every six or seven years now do more that make things a little dry; they lower the water table another ten feet at a crack—forever.

The hydroelectric power from the Sierra dams nearby has long since failed to sate our Valley's electrical appetite. Our planners, once thwarted, still talk grandly of coal plants (with new "scrubbers," of course). The coal is to be mined in Utah and sent rolling in here each morning. (The mountains on each side of this very level valley will offer their good acoustics for a constant roar of hundred-car trains.) At the moment, people on the coast—ocean breezes blow their fossil smoke toward us—play Athens to our Thebes, and still block power-plant construction.

In short, we have come full circle. If, in 1850, our Valley was untapped and useless for farming, so too it will be again by 2050, when the land, water, and air are exhausted. There are a few places still—one small tract lies on our own farm—that have never been cultivated and thus serve as museum pieces of the natural landscape. Such ground is, of course, ugly. It is full of weeds and wild willow, and by May is no more than scorched earth. The local ecologist tends to describe this type of natural trash heap as "a rare pristine ecosystem full of indigenous grasses and insects." That blot does offer a radical contrast to the verdant trees and vines that surround it. Despite what the university pundit says, the quarter acre of unspoiled waste is nature's wild fraternal twin to the Edge City growing along the 99 a mile and a half from my farm. There we have torn up vineyards and planted the following: Wal-mart, Burger King, Baskin & Robbins, Wendy's, Payless, Andersen's Pea Soup, Holiday Inn, McDonald's, Carl's Jr., Taco Bell, four gas stations, two video stores, a car wash, and three shopping centers— an overnight emporopolis, a modern caravan stop to serve a hundred thousand drivers, 24 hours a day. Its neon, smoke and noise are an Antioch or Tyre, a reminder, like our section of scrub, that both man and nature in their extreme can be altogether ugly.

At best, agriculture provides the fragile balance between the neglect and the ruin of nature. It serves at the mean, the rare equilibrium between the work of the mind and the labor of the back, where man can cultivate the wild and neither destroy it nor be destroyed by it.

God, forgive us for what we have done to this Valley, although we know what we have done. And grant, too, that our madness stop short of the utter destruction of this precious oasis. Allow sanity to return to your children before they go on to ruin what others, older and better, once created with you guidance. ▫

Victor Davis Hanson is a senior fellow at the Hoover Institution at Stanford and farms in Selma. His most recent book is Ripples of Battle: How Wars of the Past Still Determine How We Fight, How We Live, and How We Think *(Doubleday). Website: www.victorhanson.com*

1974

Phyl Diri

*I*n the years since my childhood in the fifties, the farms at the north end of the San Fernando Valley had given way to subdivisions and the eastern part of the valley had begun to slough into urban decay. The eastern end of every city in California, it seemed, had become poorer, and the western ends richer, as if there were a natural force creating the distribution of wealth.

When I was eleven—in 1957—I had bicycled from our home in Silverlake to Highland Park and passed blond hills where firemen stood burning the grasses black. In 1974, when I returned to L.A., young men drove by the old houses and shot into them. The boys laughed when they saw families inside the houses drop to the ground behind their picture windows.

In the sixties, the city took the sweet rural village of Chavez Ravine through the power of eminent domain and gave it to Walter O'Malley as an inducement for him to bring out the Dodgers.

The city also took the Victorian houses on Bunker Hill. Old women—their cheeks rouged in magenta circles, their eyebrows penciled into black arcs—used to stand outside in the shade on Sundays wearing Sunday dresses and carrying patent leather Sunday purses. In 1974, the hill was bald dirt except for the phallic thrust of one multi-story bank building and a multi-level parking lot near the courthouse. The sweetly named funicular railway, Angel's Flight, remained, and its cars crawled up and down the hill as they had for as long as I knew: the descent of one car pulled the other up.

While I had been away, my parents had moved from Silverlake to Los Feliz, not far from Griffith Park. In the way that the farm owned by a Dutch family, the Broncks, had become the Bronx, so the rancho owned by the Feliz family had become Los Feliz.

My parents' new house was larger than the one in Silverlake.

The new neighbors were richer, although the very wealthy lived in mansions behind gates on Los Feliz Boulevard and in smaller houses on the lower hills of the Santa Monica Mountains.

The neighbors in the big houses sat in chaise longues by their pools and watched brown men from Guatemala rake the violet jacaranda blossoms from the drives and cut back the red hibiscus and bottle-brush plants and toyons and the blue plumbago from the paths inside the gates. Sprinklers iced the air with tinsel water over delphiniums and hydrangeas and Midnight Blue roses. "Lazy people," the neighbors said to each other. The neighbors rose and walked into their big clean kitchens with pine cabinets to hold dishes. They watched the brown or black women scrub the stainless-steel sinks. "Dirty people," they said. The white children came downstairs wearing the clothes washed and ironed by those women.

The brown workers went home to guitars played on porches, to aunts and uncles at birthdays, to Gordito with his round head learning how to walk, to their daughters' lacy white confirmation dresses, to welcoming new friends with a handshake and old ones with an embrace. The black women boarded buses for the long trip to Watts, a large neighborhood of small houses with corn and tomatoes growing in front gardens.

Los Feliz had apartment buildings—some lovely with French doors leading to gardens, high ceilings, arched doorways, and parquet floors. Rents were still low, although I couldn't find a place because I was a divorced woman with two children. After eight months, I found a tiny cottage in Hampstead Heath, the only area of urban blight in floral Los Feliz.

There had never been a real heath in Hampstead Heath. Wild grasses had once grown the color of straw, and light through the olivaceous leaves of the live oaks had once cast deep shadows, but that had been before the Anglos arrived. Silent-screen actors lived in the Heath when Barnsdale Park was Olive Hill, before Frank Lloyd Wright designed the Huckleberry House on the hill, and when Hollywood Boulevard was a dirt road called Prospect.

The onsite manager lived in the apartment building next to the Heath. She was middle-aged and wore a different wig every day. Some days, she was Marilyn Monroe. On others, she was Jackie Kennedy. That year was a year of hair. Middle-aged women wore wigs and falls made of artificial hair. Men wore beards and sideburns. The wigs sat on the bald pink heads of mannequins in the window of Woolworth's in the Barnsdale shopping center. One head was brown and wore an Afro and looked like Angela Davis.

"I'm taking a chance on you," the manager said when I handed her the rental application. "You've got kids. I'm the only manager around here who lets in kids." She showed me a photograph in a silver frame of a young man.

"This is why," she said. "He's my son. He's my life. My husband left me and no one would take us after that until I got this job. It was so hard. Do you ever wonder, 'Why does it have to be so hard?'"

You have to believe something can be different than it is to answer a question like that. I did not. I carried my difficulty on my back like a snail carries its shell. I knew that much.

The Hampstead cottages had flat tops rather than true roofs. If you were to look down at the roofs from a helicopter, you would see heavy gray paper crookedly sealed with snakes of black tar. Small windows went around three sides of the houses. Over each set of concrete steps in front rested very small beaked roofs sticking out over the stairs. Except for the little beaked roofs, the cottages stood square and without ornament.

One wall contained a large square ridge for a Murphy bed that pulled down. When the Heath was built, there may have been ice boxes, but not at the Heath; built into the kitchen was a cabinet with wire racks for cooling. Our refrigerator rested in the living room, and occupied much of the room. The ancient water heater was illegal and dangerous, and sediment thumped ominously along copper coils. The gas company sent a warning notice to the management company, and the company replaced it, but, when I phoned to tell them my neighbors also had illegal water heaters and that if one blew up all of us would blow up, the man in the central office said, "You want trouble? Sounds like you want trouble. Move." The gas man told me we would die if we used the wall heater.

Nonetheless, our poverty felt almost harmless. It was in the social interstices of Los Angeles that the abandoned, those in flight, those trying to get work in the studios, the schizophrenics, and the revolutionaries lived. Almost anyone could become almost anyone else. One might not be poor forever. The Vietnam war had ended; anything could happen.

From my mother's point of view, however, I had reversed my family's trajectory of escape. My poverty embarrassed her.

"Remarry," she said. "Marry a rich man." If I had not chosen a husband because he danced well, she meant, I would not have been a failure. For my mother, a woman's only work was marriage. She looked around our tiny cottage with disgust. It reminded her of the tenements of her past. My father was not a rich man, but he was not a poor one. If he also felt I failed, he didn't say. He was not a big talker.

An old palm grew in front of The Heath like a callused gray elephant foot coming apart around the toes. Past the back of the court ran two thin bands of concrete to the garages with wooden doors hinged to open for the narrow automobiles of the 1910s.

The cottage near the palm tree stood empty by day. At night, twenty farm workers and their children lived in it. The day-workers, the Barnsdale bag lady, and the burglar who lived in one of the garages, all used the pedestrian underpass as their latrine. It gave off uriniferous fumes.

Weird Bob lived in one of the four-plex cottages near the garages. He had the profile of a Roman senator. He combed thinning hair back from a noble forehead. Weird Bob worked for the studios and won a Oscar for his work creating giant photos that lost none of their detail; he lived on a diet of cocaine, beer, and vitamins. He came from Ukiah, the county seat of Mendocino.

Weird Bob had one of the two cars in the Heath, a rhinoceros-colored Auto da Fé with push buttons instead of a gear shift. An electric cord hung from the engine, and the cord ended in a plug. The car was so formidable it looked as if it should have gun turrets on its sides.

Paranoid Bev lived in the four-plex in a bottom unit. She was the red-haired boy's mother, a house cleaner, and a poet. She often wore a bright skirt, a fringed shawl, and a hat made from a lamp shade with the word "Tijuana" embroidered in script along its bottom.

Paranoid Bev believed her gas bills revealed the conspiracies designed against her by her husband. We never saw her husband. For all we knew, he lived in the claw-footed bathtub behind the shower curtain. She asked me to review her gas bills, and I did.

She stood watching me, sneering at what seemed to her to be her husband's manifest trickery. The pieces of paper were just regular gas bills. I suggested she stop using her wall heater. I thought the carbon monoxide disturbed her mind. She said, "Yes!

That's how he's trying to get me!"

Paranoid Bev crammed her cottage with things she found: three televisions that did not work, dozens of little plastic cowboys on horses, a bronze bust of Milton Friedman, a framed copy of Rudyard Kipling's *Advice to A Young Man*, a lamp base depicting Prometheus bound to the Caucasus Mountains, boxes of free clothing she hoped would fit her son but did not, and, once, a branch that had fallen from a tree.

Berto and Peggy lived in the cottage on the other side of the driveways. He was a young black man, and he sang with a rock 'n' roll band. She was a beautiful white girl. They left Ohio because Peggy's father found out she was in love with Berto and tied him to a tree and whipped him. They had planned on living on the fruit of avocado and orange trees. The Heath had one avocado tree. Berto and Peggy stripped its branches of even the smallest avocados. They were very hungry.

When Berto had a gig, the young couple bought liver and placed it on top of their water heater to cook. After dinner, they ran out from the cottage with joy and led the Heath children in a line dance. Berto sang "Rocking Pneumonia and the Boogie Woogie Blues" in his sweet contralto.

When he did not work, I invited them to dinner every night. They offered to pay me back by applying putty to the windows of my cottage to keep the wind out, and I accepted.

In the cottage beneath Weird Bob's lived a psychology student, a large white girl who stored bacon grease in her kitchen drawer and taught the Heath children to spit so as to make a straight line above her sofa. Her name was also Bev. The children called her Heavy Bevy.

Opposite me lived a middle-aged couple who collected coins from the laundromat on Vermont, and next to them lived their adult daughter. A young anarchist husband and wife lived with their little girl in the house on the other side of the Laundromat people. They organized the Hollywood Cooperating Community down on Hollywood Boulevard in the basement of an office building. They read Marx together in the shade of the elephant-like palm.

Further down, opposite the day workers, lived an Expressionist artist who painted wonderful pictures full of emotion when she was all right and tried to kill herself when she was not. I kept a glass jar in my living room to save quarters for gas for the Auto da Fé to take her to Kaiser ER. She was one of Weird Bob's lovers. His other lover painted giant toothpaste tubes and made fruit and vegetable costumes for dancers in commercials. She told me that, on an acid trip, she learned she was Jesus Christ with tits, so we called her The Woman Jesus. She lived next to my children and me.

The Chicano children from the crowded end-unit were all boys. They played with the red-haired boy on the rectangle of lawn between the cottages, although they did not like him. My daughters did not want to be around the red-haired boy. He left drawings in our mail box of naked women with targets on their genitals.

My mother hated the Heath. "These people," she said, "do not wear deodorants." She drove down, however, to visit her grandchildren. The Heath was only a mile from my parents' house, but she could not walk down in high heels. She wore nice knit dresses and sprayed her hair and put on red lipstick and face powder and marched through the wall of stench from the underpass as if she were the Queen Mother on a visit to the underprivileged. She always arrived unannounced.

On the last visit she made to the Heath, my daughters and I were across the street at the shopping center buying groceries. Berto was at my house, slowly applying putty at the base of a pane of glass. He looked up from his work. He had already made many sets of tiny false teeth in the putty, and he was just then polishing a small putty sculpture he called the "Ficking Licking Chicken," a chicken in a Jane Russell bondage bra. My mother saw him and demanded to know what he was doing to her daughter's house.

He removed a joint of the Hawaiian marijuana from his lips and peered at her myopically. "Mama?" he asked vaguely.

"Lady," Heavy Bevy called from her porch. "Your face will freeze like that."

My mother backed up abruptly and ran into Weird Bob. She turned fast and looked him up and down. "Hey," Bob said softly, "who starched you?" She made her retreat with her back stiffly erect, taking umbrage as she went.

Paranoid Bev also watched from her porch. She turned to Weird Bob.

"I bought a gun to kill Berto. He pushed over my little boy."

"When was that?"

"In June, when he wore the Santa Claus suit."

"No, Bev. Berto didn't."

She shrugged. She knew better. "I'm going to invite Berto to my place and shoot him. I will tell the police he tried to rape me, and the police will believe me. White people think black people are strange," she said knowingly.

Bob looked at her. She wore the Tijuana lamp hat, candy-cane striped stockings, an orange-and-pink flowered muu-muu, and a string of white pukka beads.

"You had the police here three times last week because of a noise," he said. "They won't believe you. Don't do it."

"That was different. How was I to know there was an asthmatic rabbit in the bushes?"

"There was no fucking rabbit, Bev." She stared him down, so he went upstairs and lifted weights.

A short time later, Bob heard a shot and the sound of breaking glass. "Shit," he said and lowered his bar bells. An instant later, he heard Bev's screen door slam and then the pounding of feet on the pavement. He looked from his window: Berto ran zigzag over the grass rectangle. He jumped in his van and, after a few frantic attempts, started its engine.

He saw my daughters and me walking toward him, and he leaned out of the window of his van.

"She tried to shoot me! She has a revolver and she shot out one of her windows!" We did not have to ask who he meant. We opened the side door and got in with him and headed for the police station.

Berto drove in concentric circles, circles encompassing greater and greater areas of Los Angeles and its vicinity, and he sang increasingly lovely songs about a wino who played an illuminated glass saxophone. We ended up in La Cañada watching a "Gilligan's Island" marathon on a black-and-white television set at the apartment of a man called Tony the Dope Dealer.

While we were away, the red-haired boy bought a purse-sized mirror from Woolworth's. He stood on the concrete platform where the garages used to be before they burned down, and he flashed the mirror at the police helicopter.

By the time we came home, the red-haired boy lay flat on his stomach on the sidewalk with his arms and legs stretched out. Cops ran their hands along the boy's legs and back as if he were a grown-up criminal. The Heath was silent. It was always silent when the police arrived. My neighbors closed their doors and drew their shades. Fourteen toilets flushed.

We held a party in every house but Bev's that night. Her cottage was dark except for the light from her flashlight as she reviewed her gas bills. A member of the Platters singing group came, a friend of Berto's. The man wore leather pants and a silver-and-platinum-and-diamond bracelet that looked like a waterfall flowing over his dark wrist. The large psychology student wore her bikini. The Chicano students flashed little mirrors up at the sky. The Woman Jesus and the Expressionist painter danced a solemn dance together in one corner of the grass rectangle. The bag lady in her maroon velvet hooded cape stood by her shopping cart, waving her arms over her head, dancing to an interior seaweed music. Weird Bob had blown up giant pictures of Berto's face and pasted one face in on each of Paranoid Bev's windows. Before her screaming stopped, he ran around and removed them all.

The Heath era went on a little longer. Tall palms punctuated the sky. At dusk, sidewalks around the Heath were planes of gold.

Then Berto moved to a studio apartment in Hollywood and sold men's jewelry at the Broadway. Weird Bob got a job teaching film at UCLA. After he contracted encephalitis, he moved back up to Ukiah. The Woman Jesus moved to the Santa Cruz mountains and lived in a small cabin behind a family from Russia.

I bought the Auto da Fé for $250, and the anarchists loaded up the truck from the Cooperating Community with my furniture, and my daughters and I moved to West Los Angeles, where I went to law school. I never learned the purpose of the electric cord.

The California Supreme Court held that tenants have a legal right to decent housing. Seven years later, the Court found that landlords who refused to rent to families with children violated the Unruh Act.

In a torrential storm in 1981, the police found Tony the Drug Dealer's body in the trunk of his car at the bottom of a ravine in the Topanga Canyon. His family buried him on a brilliantly sunny day, when the hills around the city grew vivid green grass and tiny purple wildflowers.

By 1990, the poor spent almost everything they earned on housing, and the interstices of possibility diminished. The Pacific became contaminated with mercury near the coast, and the sky over the ocean near Santa Monica was often brown.

The wealthy neighbors in Los Feliz grew old, and no one listened to them when they looked around, saw that some of their neighbors were people of color, and said, "Why, they're taking over."

My parents moved to Palm Springs. Twenty years later, my mother died, and I cried too hard to be able to speak. A priest spoke and gave me a crucifix. I looked up at my father. He cleared his throat. "You never know," he said. "She might have been right." A man dressed to look like Elvis Presley took my empty hand and said in a deep voice, "I'm so sorry for your loss," and I realized that even that terrible day contained the two notes of tragedy and nonsense that sounded my life's theme.

My father lived alone after that. Every time I went to Palm Springs to see him, he said very little. We went to the cemetery and placed flowers he took from the neighbors' yards in the metal cup near my mother's grave marker. He talked with her for a few minutes, and we got into his car. "Do you know what life is? I'll tell you." He said this each time we drove away towards the San Jacinto Mountains. "Life is memories. You have to make sure they are good ones."

My daughters went to college, married, and had children. They saw the red-haired boy two years ago in Peet's Coffee in Berkeley. He was bald and he wrote books on nature for the Sierra Club. They were glad to see each other again. They know that all memories are good ones, even the bad ones. If the memories are not good, you rewrite them. □

Phyl Diri is an attorney who lives in Salinas.

THE BALLS OF MALTA

L.T. Jordan

*Ireland, Sir, for good or evil is like no other place
under Heaven, and no man can touch its sod or
breathe its air without becoming better or worse.*

George Bernard Shaw

*I*t was, as they say at home, a nice soft morning as I made my way up Grafton Street to meet my intended brother-in-law for a couple of pints and a chat.

We call it "soft" whenever the rain falls horizontally and we hope to God it will pass quickly.

So there I was, rain streaking down the back of my neck, shoes leaking, no raincoat between me and the elements. Not that I'd forgotten, you understand, I merely didn't own one. I did, however, own a heavy Crombie overcoat, and a lovely thing it was. Thrown over the end of your bed on a winter's night it was a darlin' thing. Good enough to dress out a corpse. Regrettably, it was in the pawn. Very changeable weather, as my Mother would have it, you wouldn't know what to be pawning. The fact that it had been placed in the pawn earlier that morning meant that I had a few shillings to spend, and sure, who can have everything?

Out of the downpour I ran and into the dark comfort of McDaid's pub in Harry Street there to find Lorcan, my intended brother-in-law.

"The hard man, Larry. How's the form?"

"Not bad, Lorcan. Yourself?"

"Grand, thank God. Eh, Larry, this is Brendan Behan."

I should mention that this conversation, like many others that followed it, took place entirely in Irish, for Lorcan placed no value on things English, nor did Brendan who had gone on record at the age of 16 by trying, unsuccessfully, to blow up a British warship in Liverpool.

He had gone to Borstal for his trouble and while there had taken a leaf from another Irish writer who, upon reading *War and*

Peace, announced, "Jaysus, but that's a grand ould story. I think I'll write one meself."

Brendan's first effort, "I Was a Borstal Boy," appeared in an Irish journal called *The Bell* in 1942. Much later he turned this material into *The Borstal Boy,* which became a bestseller and prompted Kenneth Tynan to write: "If the English hoard words like misers, the Irish spend them like sailors; and Brendan Behan, Dublin's obstreperous poet-playwright, is one of the biggest spenders in this line since the young Sean O'Casey. Behan sends language out on a swaggering spree, ribald, flushed, and spoiling for a fight."

Dublin, as someone else once said, is a city where wit is prized above riches. It is also one of the most written about and least cared for cities in the world. A city which sometimes manifests a curious and somewhat ineradicable bent for the second rate, as though it found the first-rate rather uncomfortable to live with. Or simply boring.

On one of my rare visits to Dublin a few years back—notice I didn't say holiday, for a trip to Dublin for me is more of a pilgrimage than anything else—I spent a few quiet hours in a pub with Brendan's widow, Beatrice. Oddly enough, it was raining that evening, too. We chatted pleasantly about any number of things, including Brendan, his work, his friends, and the like.

Her own story, *My Life with Brendan,* had been published the year before. It is a sad story of a man who could quote an 18th-century Gaelic poet as readily as he could quote Joyce, Yeats, O'Casey, or Wilde. A story of the man who said of himself, "Success is damn near killing me. If I had my way, I should prescribe that success go to every man for a month; then he should be given a pension and forgotten."

She felt, she said, that she had to write it. She had been quite upset with Ulick O'Connor's recent book, *Brendan,* and its allegations about Brendan's supposed bisexuality.

When I asked her what life was like now, she answered, "It's like living in a railway station."

As we approached closing time, I asked how she thought Brendan would like best to be remembered. Her answer took me off guard, and yet, upon reflection, it was dead on the mark. "Brendan would like to have a theatre named after him. That would be nice," she said, and her eyes looked at me with the sadness of vanishing light.

But I'm getting ahead of myself.

The morning in question was in the early fifties, 1951, I think, and Brendan was still earning his living as a housepainter. His literary career, like his death in 1964 at the age of 41, was still in front of him.

My intended brother-in-law was at that time an unpublished poet who spent most of his working days balancing mathematical equations in a government section called CIE, for Coras Iompair Eireann, or the public conveyance department.

Like most government bodies, CIE never seemed to make a profit, and given the irregular schedule of their buses, I'm not surprised. Listening to politicos forever justifying the economics of it all would put bloody years on a man. Seems like we used to stand at a bus stop for a lifetime waiting for a bus to come along. On one occasion, I heard an old woman screech at a bus conductor, "It's not CIE at all yiz should be callin' yourselves, but the bleedin' Banana Bus Company for yiz always come in Jaysus bunches."

How McDaid's became a hangout for poets, writers, actors, painters, and a couple of defrocked priests is beyond me, but if you sat there long enough you'd be worn out shaking hands. Perhaps it was because the place was an ass's roar from the Gaiety and the Olympia Theatres, or that Trinity College was down at the end of Grafton Street, or that the College of Surgeons and the National University were around the corner.

Or maybe it was because people felt comfortable in the place. It was certainly without pretension. You'd never see a policeman in it, and the pint was very good.

The barman called everyone "Mister EH," which, I suppose, made us all feel equal. No names, no pack drill. "Ah, good mornin' Mister EH. Nice mornin', thank God. Yes, it is. A pint is it? Right yeh are." What can't be cured must be endured.

"Ah, Larry, ould son, Brendan here was just telling me a gas bloody story," said Lorcan, handing me a pint. May the giving hand never falter.

"Seems like some Yank came out of the Shelbourne Hotel the other morning and got into Whacker Nolan's taxi. Poor man was devastated by the incompetence of the Celt, as he put it, and asked Whacker if we Irish had a word for *mañana* like the

Mexicans do. Whacker, God love him, never bats an eyelid and says to yer man, 'Yessir, matter of fact, we have three of them, but they all lack the urgency of the Spanish.'"

There are worse ways of spending a rainy morning, I'll tell you.

Brendan and I shared one thing in common. We were Northsiders, having the dubious distinction of being born a few short streets away from each other in the slums of the north side of Dublin. Lorcan, on the other hand, was born in Dun Laoghaire, a fashionable suburb on the south side, as the newspaper columnists would have it, although I always thought God broke the shovel when He made that place.

Brendan was about to tell another story when the door of the pub opened on an elderly and very wet postman. "Large glass of Power's, please," said the postman and dropped his cape and postbag to the floor.

"There's no such thing as a large glass of whiskey, as Oliver St. John Gogarty once remarked," said Brendan.

"Ah, yes. Yer man. How's he keepin' at all?" inquired the postman.

"The ould bollox is not well at all."

"Jaysus, all the ould crowd is droppin' like flies."

The next time I met Brendan was about a month later in a pub in Henry Street. Long since gone, it was called the Tower Bar and looked across the street at Radio Eireann, the state-run broadcasting system. We were fortunate enough in those times, I suppose, in that we didn't have to waste time choosing a particular radio station to listen to, for there was only the one, Radio Eireann.

If you had a strong aerial you could always tune in the BBC. This, however, was considered unacceptable behavior in my circle of family and friends. "God knows what class of cod's wallop those people would have you listen to," was the way my father phrased it.

The Tower was home to any number of radio types who were given to dashing across the street between broadcasts, throwing down a few quick gins, and racing back to announce the travails of mankind to anyone who might be listening. It was only strangers who lingered close to the door. Regulars knew the dangers of blocking the way of hungover announcers.

"Jaysus, Paddy, a large gin, and quick. The nerves are gone."

"Why don't yeh have them out, like the teeth?"

"Never mind the bloody chat, and give me another. Put it on the slate."

The check-cashing and bill-paying procedures in the Tower on Saturday mornings would do justice to the House of Rothschild.

Anyway, there we were on a warm Saturday morning, Brendan, Lorcan, and myself, having a few jars on the strength of Lorcan's paycheck, when we were joined by a radio actor whose name I've forgotten. He was, I recall, a large man with rabbit eyes, small soft hands, and a voice as gentle as a summer lake. He also had the reputation of being a "toucher," one who is forever borrowing money and forgetting where he got it from. But in a nice way.

"Ah, yes. The £5, of course, to be sure. Would Friday evening be time enough for you? Terribly sorry about the delay. Had it for you last Friday, but didn't see you here."

The radio actor was, as usual, in the horrors. His complexion was reminiscent of white blotting paper. "Did yeh have any breakfast at all," asked Brendan. "A few eggs and rashers would do yeh the world of good."

"Merciful Jaysus, Brendan, but I couldn't look at the flag this morning," said the actor and called for a double gin.

"I'm stuck with a bloody author this morning, if you don't mind," he continued. "Introducing his new book on my programme."

"Any author who has to introduce his own book shouldn't have bothered writing it," said Brendan.

"Understandable, Brendan," said the actor, "but he's going to read from it as well."

Brendan lifted his pint to his lips, drained the glass, and ran the back of his hand across his mouth. "Reading your own stuff," he said, "is a form of mental incest."

Shortly after that exchange Brendan shouted a greeting at a small man in a felt hat, and walked away from us. I recognized the man as Brian O'Nolan, famous columnist on *The Irish Times,* whose writings appeared under the pseudonym Myles na Gopaleen, or Myles of a few quick gins, and racing back to announce the travails of mankind to anyone who might be listening. It was only strangers who lingered close to the door. Regulars knew the dangers of blocking the way of hungover announcers.

"Jaysus, Paddy, a large gin, and quick. The nerves are gone."

"Why don't yeh have them out, like the teeth?"

"Never mind the bloody chat, and give me another. Put it on the slate."

The check-cashing and bill-paying procedures in the Tower on Saturday mornings would do justice to the House of Rothschild.

Anyway, there we were on a warm Saturday morning, Brendan, Lorcan, and myself, having a few jars on the strength of Lorcan's paycheck, when we were joined by a radio actor whose name I've forgotten. He was, I recall, a large man with rabbit eyes, small soft hands, and a voice as gentle as a summer lake. He also had the reputation of being a "toucher," one who is forever borrowing money and forgetting where he got it from. But in a nice way.

"Ah, yes. The £5, of course, to be sure. Would Friday evening be time enough for you? Terribly sorry about the delay. Had it for you last Friday, but didn't see you here."

The radio actor was, as usual, in the horrors. His complexion was reminiscent of white blotting paper. "Did yeh have any breakfast at all," asked Brendan. "A few eggs and rashers would do yeh the world of good."

"Merciful Jaysus, Brendan, but I couldn't look at the flag this morning," said the actor and called for a double gin.

"I'm stuck with a bloody author this morning, if you don't mind," he continued. "Introducing his new book on my programme."

"Any author who has to introduce his own book shouldn't have bothered writing it," said Brendan.

"Understandable, Brendan," said the actor, "but he's going to read from it as well."

Brendan lifted his pint to his lips, drained the glass, and ran the back of his hand across his mouth. "Reading your own stuff," he said, "is a form of mental incest."

Shortly after that exchange Brendan shouted a greeting at a small man in a felt hat, and walked away from us. I recognized the man as Brian O'Nolan, a famous columnist on *The Irish Times,* whose writings appeared under the pseudonym Myles na Gopaleen, or Myles of the Little Horses. He was also famous as Flann O'Brien, the author of four novels in English, as well as the famous work in Irish *An Beal Bocht,* since translated as *The Poor Mouth.* James Joyce said of O'Nolan, "A real writer with the true comic spirit." And Brendan is quoted as saying, "I read him with relief and jealousy."

When O'Nolan left the pub, Brendan returned to our company, laughing. "O'Nolan just told me he saw a huge Buick outside the American Embassy in Merrion Square, and that it reminded him of a pregnant whale blowing a mouth organ."

Thus, a wave of laughter followed O'Nolan's exit into the warm sunshine.

But our glasses were bordering on empty and I doubted if we had the price of three pints between us. "Jaysus, but a ball of malt would go down well," said Brendan, a ball of malt being what a double whiskey is called in Dublin.

Just then a young man entered the pub dressed in the resplendent gray-and-red uniform of the Knights of Malta. In his outstretched hand he held a cardboard box bedecked with small paper flags on pins, these flags bearing the Maltese cross and being sold for whatever you'd care to give, to support the charitable works of the good Knights. You slipped your money into a slot in the box and the Knight placed a flag in the lapel of your coat.

The young man looked at us and rattled the few coins in his box. "Merciful Mother," intoned the actor, "what in God's Holy Name is that?"

"With any luck at all," said Brendan, "it would be the money for the balls of Malta." □

Larry Jordan, an ardent golfer and frequent lecturer on Joyce, O'Casey, and the Abbey Theater, was born in Dublin and emigrated to Los Angeles in 1960.

ONO ONO GIRL'S HULA

Carolyn Lei-lanilau

Now that I have had one or two things published and "other" Chinese Americans are making loads of money writing fiction, the hopeful phone calls and letters from my literati relatives in Hawaii and China urge me to join ranks with the Chinese American fiction set: "Dolling, why don't you write fiction and make oodles and oodles of money? Poetry is so *hard*." And my husband, who knows what no-good I'm always up to, shakes his head in agreement: "Philosophy's a dead end. Nobody will buy it."

How can I tell my closest people that I am dysfunctional, cannot write in vertical columns?

While lots of folks like to eat Chinese food from boxes and have even seen an "about Chinese" movie, in all probability, they were only privy to whitewashed Chinese Americana. What I mean is that the average American Chinaphile is so sheltered about Chinese. Take me, for instance. I grew up "basically Chinese." Though my father is motley Chinese, I am mostly gypsy-gened Hakka. While the Chinese around me were influenced by the Hawaiians—a complete system in and of itself—my orientation was primarily bourgeois. At one time, I could speak French better and could translate Latin, but dared not utter *ching* or *chong* or *ling* or *tong*. Did not want to yell that loud grease-talk.

As a kid, I saw one or two movies. Had the classic Pearl Buck grandmother book read to me. I had seen Taiwan-Mandarin and Cantonese, but more European operas. Only when I had my kids—that's when I began to groove in Chinese. I forced my older daughter to study characters on Saturdays instead of allowing her to join Bluebirds. After Ana was born, I lied. I told Mrs. Lee that Ana was six when she was still four, just so she and Eirelan, who was nine, could have the joy of studying together. I studied Mandarin. I pushed away my friends and family for about ten years just so I could start from the beginning and maybe end up as an old lady satisfied that I had read all the literature, philosophy, and history written by Chinese. I was insane.

During that time, I went to China twice to study and work. I only had Chinese-from-China boyfriends. And I was selective: they had to have titles and exotic occupations such as "paleontologist" or "semiotician." My daughters hated me. Then, by accident, I met my husband. The one place I had vowed never to go to he's from: Xi'an. You know the story of the ballerina and the tin soldier? He's the stone soldier. The point is that the ballerina did not know military life and the soldier never knew *salon*. I am living the intertextual real stuff—the nonfiction about my "feeling" Chinese, Hawaiian, American. Those "other" Asian American accountants are so removed and outdated.

This crusade begins in Hawai'i. Not the Hawai'i with condos and shopping malls. No, that is not Hawai'i. The Hawai'i that most people know is some mutant demon that Pele is demanding due respect. When I grew up in Honolulu, nobody ever saw a real eruption except people on the Big Island. Pele was happy: no active volcanoes! We had dull happy lives in our grass shacks. We all lived in dried-grass mansions. Naked, mostly, we mumbled and grunted. We lived on seaweed and sharks' fins, and swung on vines. Life was slow. The air was still and always quiet. Distinctly, I remember reading *Moby-Dick* and wondering why sexual instincts had not yet revealed themselves in my adolescent blossoming. I looked out my bedroom window. There was the hung wash.

Eventually, my mother and I would leave to visit my bedridden grandma. My room was neatly in order. Nothing in or outside the room was misplaced, and it had to do with the stillness or the air that moved in the territorial frame. Innocence. Island mentality. Provincial upbringing. All I knew was private school in the day, speaking good English at home at night. Nobody spoke pidgin in my father's house. Nobody spoke Chinese. If you said a word in English, it was *de rigueur* to say it right, because my mother always trailed with *Correct* on her hard drive. If you mispronounced a word, she nailed you. When my father's relatives showed up from Australia—talk about fun and picking up accent! But as far as "correct speech" was concerned, nothing but logic and diagrams were acceptable to the family.

Imagine how stunned I was when I transferred to a girls' high school and overheard "fuck" spoken—quite correctly, with the hormones and the discontent towards authority active. I guess I was "sheltered." I was fascinated with these lovely hula "island" girls, swishing in their pleated skirts, lips reflecting torch gingers, plumerias, and respectfully bred hibiscus, forming the "f" like a cannon loading, then burning the smooth saintly air with the explosive "uck." Though nominally registered, this piece was yet another factor in my footnotes on "I hear. I see. I think, therefore, so what?"

The gods in the one temple were Mildred Chong and John Winfield Lau. At my other house, it was Tai Yi and five older cousins. Chez parents was like ballet class: with each spoken word corresponded a physical act. My mother's favorite terms were, "That is Cor-rect?" or "That is In-correct?" My father never uttered a sentence to me after he mourned the fact that "if you were a boy, I could take you to the baseball game." He was a god. Orderly, beautiful script. Made money. Secretive. My father was a statue, a portrait of a flag. He was a Republican and probably much more so than the guy who went to 8:00 a.m. mass every Sunday. He rarely drank and once in a great while he played his ukulele and sang Hawaiian songs. My mother, expert in worrying, never on time, always 15 to 30 minutes before schedule, compromised her deficiencies by claiming that "playing poker is my only recreation." Ideas are my "only recreation."

The "other" mother's house was much more "civilized" and culturally developed. There was a piano, a library, a phonograph. Tai Yi, my mother's eldest sister-in-charge, would now and then open up secret closets and peel out brocades, silks, silver dollars, jade. There was an altar in the kitchen and another outside on the spreading veranda under the lichee tree. There were so many secret violet and heavenly white, spidery yellow, and green-brown orchids. And mango trees and starfruit and a white peach tree and sugarcane and an avocado tree and a pomelo tree and a *pikake* bush whose branches she would occasionally break off and whip through the air. Imagine how that hurt when used on me! Everything, everything was good. The food, always yummy.

There were five cousins to babysit me—I had five different adventures per week. (Was I "spoiled" early in life?) Nelia was a daredevil, the first to own a 35mm camera and visit Europe! Egypt! The mainland! Beatrice played the piano. When I was being punished or just watching the air, she might start fooling around with Tchaikovsky and I would be instant mango slurp. I am still in love with Stanley, who promised me the moon and became the archetype for every young man thereafter. And I'm still in love with Jimmy, his younger brother, who was "the baby." Florence was a corporate exec-in-the-making. I love my five cousins who crusaded my imagination and desires and flooded me with over-expectation in this life.

One afternoon, Tai Yi began to rap in our secret Hakka language for me to run and hide in the back bedroom. She bolted the doors and windows. She sealed the drapes and lace curtains. "Get under the bed! Get in the back!" she commanded. Not me, I needed to see. What was so urgent? "Get under the bed!?" I never saw my cooking-and-washing-clothes aunt so crazy except when she was behind the wheel and somebody was cutting her off. "You goddamn sonavabitch!" she'd cheerlead, "I have *children* in this car, are you blind!?" But we were in the house, not the car. She was frantic when she saw how obstinate I was. "You brain-damaged squash! Heaven have mercy on this nitwit!" There was no time for dialogue. She pulled my collar to the floor, while we peeked through the Venetian blinds of the parlor. There was the mystery: It was a dark thing. What was it? "That," she explained, "is a black demon." What was a black demon? I saw a black man; he was black, but where was the "demon"?

Following this incident, I continued to be dim about the matter, and whenever I overheard someone refer to "niggah pits," I wondered about all the letters between A and Z. Later, I sensed that "niggah pits" and "black demon" belonged to a similar genre. However, this was mere idea, not fact. There seemed to be a number of predicate adjectives and potential synonyms, but I was lost without a Subject. Such as "po po lo." *Popolo* was a dark sweet berry, and/but po po lo also had another meaning which was "black person." But no one explained this to me—like most complicated things, I had to visualize this out and as they say in therapy—for myself. Moreover, since I had seen only one black person in my entire sheltered life while growing up in Hawaii, this came slow. Slower than air.

My family hates change. Every time I was temporarily enchanted with some narcissistic function, some Voice would define me as "just like the father's side." What did that *mean*? Books later, while on my usual family huntdown, I discovered that it was a historical footnote meant to net me with my father's "scandalous" cousin, Robert Wilcox, who led a revolt against annexation in 1889. It meant that I had WHITE blood and was capable of *poi* dog (League of Nations) unpredictability. It meant that I didn't know my place. It meant whatever it was supposed to mean for the moment. But I was no match for the Voices. What I could do was climb the lichee tree. I talked to the branches. And orchids shared secrets. I hung upside down and listened to the sky. When you twisted the arms of these *Hawaiian kine* bushes, antlers appeared instantly, so Bambi and I were longtime soul mates. It was law that I "could not go beyond the boundary of the property," and an interview was required before anyone could play with me. My world was the compound: the family and my uncontrollable mind.

Hawai'i was different before statehood, and people from Hawai'i felt righteous just like the state motto about protecting what we got and where we came from. I mean, we're not so sure about English and everything else associated with it, but we're stubborn about being Hawaiian. Everybody that flew in after 1959 failed orientation. I'm just saying this because people go to my hometown and say, "I luv Hawaii," like they are practicing English. The Hawai'i that they are luving is the not *ono*, not good-eating buggah. That nouveau Hawai'i brought shame to us locals. Everybody says they like oldtime Hawai'i, but it is too tooounsta superfares, condo-matic, best *pakalolo,* waterfront property, Hollywood dildos and bimbos that is the new artifice. Hawaiians— and I mean, the born-and-raised brownbloods—don't like it when some haole comes to visit and writes something about Us! Cognitive Dissonance, braahh. Why doesn't somebody rebut? Why should we! Not Hawaiian style.

My mother confusingly stares at me and wonders, "Why, why did somebody so smart as you write such lousy English? You spent $3000 on that computer and what?" And I still have that Hawai'i *Ponoi* mentality. You can tell by my blabbermouth style that I'm an American and I value my passport, but I don't know or feel anything about the "rockets' red glare" or the "bombs bursting in air." I know "King Kamehameha, the conqueror of the islands, became a famous hero one day." I know the "Hawaiian War Chant." My friend, Neaulani—the prettiest, smartest, nicest person—died as a very young woman. Not any conventional death: she threw herself over the same pali that Kamehameha pushed his enemies down one by one. Neau's mother wrote my father's favorite song, *"Pua Mana."* My sister danced with Iolani Luahine, but then the hula was underground, because Hawai'i was in post-statehood shock. We were ashamed. We were backward. *Haole* complained day and night about "bad service." All the cousins from my Chinese grandma's twelve kids have made the bucks: kids going to Punahou, Iolani, St. Andrew's Priory; big houses, real estate-nobody speaks Chinese, knows anything Chinese or desires any Chinese culture. They are busy making money: buying things so they and progeny can live just like folks on the mainland. Meanwhile, haoles keep coming to Hawai'i, buying the land, building big security gates to live imitation local lives.

My husband described my living aunts and uncles as the Supreme Court. Another day, he referred to them as the dinosaurs. My aunties are exercising to keep the universe in shape. Mildred, the mathematician, does daily *wu shu* and drives her sedan just to keep alert. They all know that when they die, nobody will go to the temple. Nobody will know how to order at the restaurants. Nobody will speak our funky secret language again. We will be out of stories because we assimilated. And my Hawaiian cousins *auwe*, we have trashed and cleaned the desktop on that folder long ago. Booo. □

Carolyn Lau lives in Oakland. This piece was incorporated into her most recent book, Ono Ono Girl's Hula (University of Wisconsin Press).

Burn, Bridges, Burn

Diana O'Hehir

I have an early, giddy memory of myself as a Berkeley radical. A very young one. In fact, I am only six. And declaiming my own poetry to an audience consisting of my father and "the university couple," a department head and his wife, with whom he and I lived. (My father is a widower at this time; I'm motherless.) My father leans forward, writing something on his knee. He's taking down my poetic words as I say them, transcribing neatly onto an index card. This sounds like a latter-day fantasy, but I have the index card to prove it. This card turned up in a box of Daddy's memorabilia after he died.

My radical poetic words have to do with a homeless kitten whose body we have found in our garden, its throat ripped out. My poem blames this tragedy on the cold-heartedness of the general human world.

This poetry-reading scene happened in North Berkeley in the early thirties, in a typical North Berkeley house with "character" (timbers, plaster, fake Norman architecture, steep garden, winding paths. View of the Bay). The scene-setters are liberal, artistic people with a slight socialist bent. They are worldly, sophisticated, full of chat about art, progress, poetry—all this is fixed in my mind by loss, that frantic fixative, for my father and I were soon to leave the witty, indulgent academic couple and move across town to Berkeley's south side, an area called Claremont.

The Claremont of those days was to me stodgy, flat, and glum, as was our stucco house with its cramped front porch and no view, ditto my small bedroom staring directly into the bedroom of the house next door. But, most important, there was a new person in this house. My stepmother, who had picked this house as her bridal home. A thin, wary, depressed stepmother, too young, too tired, stiff and awkward. In spite of a well-heeled background, she looked exactly like the woman in "American Gothic." And she wasn't at all interested in transcribing my poems onto index cards.

I was by this time eight years old and already well-launched on my career of becoming a rebel, or, alternatively, an insurgent in search of escape, having acquired, as ingredients of my life, three of the pressures conducive to rebellion: an ever-present, overwhelming sense of loss; a drive toward resistance to ordinary life; and a fiercely hungry desire to escape that life.

Of course, my general sense of loss began, powerfully and ineradicably, with the loss of my mother, who was killed in an automobile accident in Prince Frederick, Maryland, where we were visiting her family. I was four years old. I was sitting on her lap. ("And with all that you expect to be normal?" a boyfriend would later ask me.) And now, four years later, I had also lost North Berkeley, with the artistic inquiry and glamour I thought it offered.

My stepmother accelerated my sense of loss. She celebrated our arrival in Claremont with a punishment that still seems special. The people next door, in the house into whose bedroom window my own bedroom window stared, had invited us all to tea. I, a bookish and fairly good little girl, lay on a chaise longue and read while the adults chatted. After we returned to our house, my stepmother announced that I was to be punished. I would be sent to bed for the rest of the day. Why? Because, stretched out on the chaise longue, knees in the air, I'd allowed people to see my underpants.

I think my father objected to this punishment, but I'm not really sure.

Even I understood that something strange was involved here. In bed, I told my pillow that I hated her; I would always hate her. I cried. I choked. I doubled my fists. I swore vengeance. Then I winched a book off the bookcase shelf and read under the covers until dark. The book was *Editha's Burglar,* a 90-page drama, which told the story of a child who befriended a burglar. I knew my stepmother would hate the burglar and would punish the child who helped him. I found the story comforting.

The Berkeley that shaped my girlhood contained actual, living writers, at least one on the verge of becoming famous. This writer, George R. Stewart, was a kind, generous, funny man who had known Jack London and who knew John Steinbeck and who exchanged manuscripts with John Dos Passos.

Berkeley also had poets, one of whom, a woman with a double name, was a friend of the university couple with whom we had lived. She singled me out at a party—in this memory I am about ten and am visiting "across campus." She bent over me, trailing a length of iridescent turquoise scarf, called me "my dear," and announced that she would have given everything she ever possessed to have had hair the color of mine (red).

There was another poet, named Ruth, who, while my father and I were still living with the university couple, wrote for me my own rhymed story of escape and flight. The characters in this story, of whom I am one, grow wings and fly off to various Oz-like places. Ruth illustrated this poem herself.

Almost all the friends of the university couple wrote. They taught at UC, and their primary writing was about Shakespeare, Blake, and Marvell. But all of them wanted to be novelists, too, and several were to publish a single novel apiece, books that wouldn't cause a critical stir, but would inspire conversation, congratulations, jokes, ironic remarks among the friends. (I am imagining this, for the novels were published after I'd been transported to Claremont; my father and stepmother didn't attend the publication parties; I couldn't sit in a living room corner and eavesdrop.)

These novels were works of social criticism, stories which attacked the status quo, elevated the proletariat. That was the kind of book you wrote during the thirties, if you were a university intellectual.

And my own father also wanted to be a novelist. The outline of a novel and some chapters were among the papers that turned up after his death. This novel is set in the midwest. It's spiky and harsh, not at all what I would have expected of him; its meticulous, class-conscious realism resembles Theodore Dreiser's.

Politics, which were exciting, sometimes intervened, although never actually in Claremont. They happened in the world beyond and were reported on by the Berkeley *Daily Gazette,* a small newspaper tossed up onto the stucco front porch every evening. A general strike was called in San Francisco. My stepmother bought canned goods and said she hoped *They* would not cross the Bay to attack our house. I spent a lot of time doing dishes and planning my escape to San Francisco to join the picket line. The strike photographs didn't show any women on those picket lines, only rows of marching men, most of whom wore caps. But I remembered Joan of Arc and thought perhaps I could make a place for myself.

My friend Eleanor and I had decided on the East Coast as a place to aim toward when we reached 18. We weren't sophisticated and didn't know any suitable radicals, socialists, or rebels on the East Coast, but, I told myself, 18 was grown-up. I could search for myself now. But Washington, DC isn't the exciting environment I had imagined; it's gray and glum and unyielding. Work is hard to find. I get a room at the Y and capture a series of menial jobs. I'm fired from one for complaining about 20-minute lunch breaks. I get fired from Annie's Quik-Stop Restaurant for pointing out that the salads are wilted.

Living takes up all my time, and it's a whole year and a half before I get anywhere near my goal of learning how to fight for the good. And that happens only through Eleanor, who has finally arrived to be my roommate.

Eleanor is here because of the war. Everything now is "because of the war," and everything is different. Washington has stopped being stolid; it's now a busy, crowded city whose employers smile at you when you come through their doors job-hunting.

I am working now at Brentano's Bookstore, and Ellie is a war bride, her husband in camp. We share an apartment on the top floor of a rusty brick house in northwest Washington. The flat is seedy. Its veneer furniture smells of cat pee, and the kitchen is a hotplate in the bathroom, but I'm fond of my first apartment.

Eleanor and I are in our kitchen corner now, making cheese blintzes. Something new to us, cheese blintzes; we'd never heard of them on the West Coast. The recipe comes from the liberal newspaper *PM.*

"Put some butter in the frying pan," Eleanor tells me.

I scrape off a generous plop of butter. "Listen, Ellie." (I'm more or less thinking out loud, continuing an earlier discussion.) "There's got to be more. I know there's more. I've been hunting for it, and I don't find it. I read that newspaper—I point at the now milk-and-flour-spattered *PM*—"they talk about unions, organizing. People meeting. Things happening, books, ideas, workers striking."

"You're not supposed to strike now," Eleanor says.

I ignore this. "Action," I say. "Remember? That's why I came East. To get involved. Be useful." Eleanor, I suspect, knows more about this possibility than I do. She and her husband were economics students back in Berkeley. One of the professors had a Reputation; he was supposed to be a firebrand. But I left Berkeley before I could take any econ. "I've been here for a year and a half, and I've been flailing. All I've done is work at dumb jobs and get fired."

"You were staying alive," Eleanor reminds me. "Jobs were hard to find."

"Yes," I agree, "they were hard to find."

Blintz-making gets interesting. "They're so good, how did we ever live without them?" Our philosophical discussion gets sidetracked for the moment.

But after dinner, while Ellie is washing and I'm drying, she says, "Diana, know what? In San Francisco, there was a Workers' Bookstore, maybe they have one here."

As soon as she says this, I know she's right. I've seen that bookstore. It's new this year, and it's across the street from the Lincoln Bank, whose sign has a neon Lincoln profile as its logo. When I saw that Workers' Bookstore sign, I thought, Hey, there's something to check into. And then I didn't. Being fired twice takes up an awful lot of your time and energy.

"We'll go together on Monday," I say, and Ellie agrees.

But on Monday, after I leave work, she's tied up, and I walk down to the Workers' Bookstore by myself.

I climb the stairs feeling pleased with the way they resound under my feet and the way the holes in the iron fretwork give you a view of the side of the building as you climb or else give you little slivers of the street below. The staircase echoes and arcs pleasantly and seems like a bridge to something new.

Inside, the Bookstore doesn't look anything like the Brentano's Bookstore where I'm working now. In Brentano's, we're surrounded by polite mahogany walls and polite perfumey new-book smells and lady customers who wear pale pink crêpe-de-Chine blouses. The Workers' Bookstore smells of coffee and cigarettes and ink. The people are mostly men in leather jackets. The furnishings are pine or gray metal shelves with tipsy stacks of books and magazines and tables and racks with more stacks of magazines, books, and pamphlets. And lots of posters, mostly colored red. One shows a muscled, kneeling worker holding up the globe; one has a woman war-worker wearing a kerchief and flourishing a wrench; another has a defense worker doing something to a turbine; a green-and-black poster shows two men in overalls, one Negro, one white, arms around each other's shoulders, with the legend: BLACK AND WHITE, LABOR UNITE.

Chairs are set up in haphazard rows in front of a podium; a meeting is happening. There's a young, khaki-uniformed, tow-headed speaker who looks like an almost-grown-up version of my brother. Tom Delaney, a sign beside him announces, District Labor Council. (And PRIVATE DELANEY NOW! adds a red-lettered paper clipped to the board.)

I find a chair and sit. Tom Delaney is talking about labor. About the working class. I think maybe he started his speech talking about labor's contribution to the war effort, but he's on to labor alone now. "We fight for the whole human race," he says. His eyes shine. "To make life better for everybody. For all human beings, all over the globe. We'll help abolish Fascism, do away with that evil monster there, in Nazi Germany. And here, brothers and sisters. Especially here, in our own country, that we love and will make better." He hits that hard.

I like what he says. I like being addressed as "brothers and sisters." He mentions the South, the mines, the Detroit auto shops, the poll tax that keeps Negro people from voting. ("Our Negro brothers and sisters," is the way he puts it.) A living wage for every American worker, he says. "And especially the women," he adds, sounding fired up about that. Perhaps he's looking at me. There aren't many women here.

"Let me tell you about myself." And Tom Delaney gets really warmed up. "I'm a government worker now. I sit behind a desk. I rent a nice apartment, well, I did all that until last week when I joined up." He points at his uniformed chest; there's laughter. "But, hey, I didn't grow up in this sort of surrounding, I grew up in a dark little town in West Virginia, a mining town. You worked for the mine, or you washed shirts for the guys that worked for the mine. My daddy was a miner. He went down in that dark hole every workday of his life, and he died a young man. My dad died at 44, brothers and sisters, though in those days 44 wasn't young anymore, because they didn't have those controls they got now.

My dad breathed in that coal dust, breathed it down into his lungs. He breathed it and choked on it and coughed and lay in bed for a year and then left my mother a widow with seven kids." Tom pauses. "Well, the union is there now, and the union is guaranteeing less coal dust and fewer widows and more life. So no workingman ever again has to cough his lungs out the way my dad did. There's a future ahead of us. Everybody strong. Everybody together." Tom raises a clenched fist for a minute, then sits down. There's pretty good applause for him.

A pie plate is passed to collect money for the District Council Organizing Drive; I put in $2.

Afterward, I go up to talk. Tom Delaney is nice. Bashful, and young, about my age. I ask if there's any way I can join a union. I feel silly asking this; I ought to know the answer. I don't think there's a union for me; I'm not a worker. But he seems really pleased when I say I'm at Brentano's. "I've got a card for that," he says. "The Office and Professional Workers. They'll be real pleased to see you."

I think maybe Tom Delaney wants to ask me out for coffee later, but a girlfriend has appeared and is hanging on his shoulder.

I go home with the union card and a big stack of books, including a poetry anthology, *Awake, Awake!* I'm reciting to myself a poem I've gotten out of the anthology:

Mix prudence with my ashes,
Write caution on my urn,
While life foams and flashes
Burn, bridges, burn.

It's a Friday morning, payday, when I try to organize a protest at Brentano's. The Brentano's ladies—the bookstore's clerks aren't employees or workers, but ladies—are standing stiffly around the coffee room, reading to themselves. They look distressed. "I simply do not understand this," says skinny, gentle Miss Tisdale, my boss in Children's Books. She hands me a slip of paper: As of July 15, we are forced to cut all five percent commissions to three per cent. This seems clear enough to me; I haven't been here long enough to get any commission at all.

"Good grief," I say.

Miss Tisdale raises her little eyebrows. "What does it mean?"

"It means they're cutting your pay, Emmy."

"Oh," she says, as if she's run a nail into her shoe.

"But," I go on, "I don't get it. There's full employment now since the war. They should feel grateful for your being here at all."

Miss Tisdale scans her paper again and sighs. "But why? Is the war cutting down on profits?"

I'd love to tell her it's because the duty of the exploiting classes is to seize the product of the workers' labor—that's the rhetoric I've been getting from my reading. I'd also like to say that white-collar workers are real workers and need workers' protection, which is what my new union believes. But all that is too hard-edged for troubled Miss Tisdale, squeezing her yellow slip as if she's wringing it out.

That night, I telephone the organizer for my new union, the United Office and Professional Workers, Local 27. Dorothy is not really an organizer, she's a social worker who does the organizing duties part-time. All the Local 27 members are either social workers or employees of other unions, except me.

"Dorothy," I say, "there's this situation." I can hear myself getting excited.

Dorothy is gentle and helpful and a good listener. You can tell she's a wonderful social worker. "Oh, my," she says, as I outline my problem. "Diana, this may be important. Let me call someone."

I review my mental list of Brentano's workers. First, poor Miss Tisdale, who is scared of me, but also likes me. And then some others that I've greeted in passing: Mrs. McManus and Mrs. Lucrece in Cookbooks, Mrs. Baccio in Travel, Dawna O'Halloran in Journals. Dawna is my age and might protest if her own toes were getting stepped on. But her toes are untouched; she doesn't get commissions, either.

Dorothy calls back the next morning, before I leave for work. "Lew is very interested."

"Who's Lew?"

Her voice lowers. "Lew is the president of the union." She means for me to be impressed, which I am.

"He'd really like for us to get in at Brentano's," Dorothy says. "Would you try to talk to people? It's wonderful to have a member in that shop.

"And I told him about you," she continues. "I said you were enthusiastic and eager. And smart."

I can't help it, I'm excited. I feel both scared and fulfilled.

"I'll really try to do this," I tell Dorothy.

I begin by persistently nudging Miss Tisdale, who seems only to half-hear me; she flutters her hands and lets one eye wander.

Next I take Dawna O'Halloran out for a frozen daiquiri. She is not receptive; she says it's the old ladies' own fault whatever happens. She wouldn't let it happen to her. She won't even be here. "I'm off, maybe the WACs. Maybe get married." I say goodbye to Dawna easily, since I don't like her.

I try chatting with Mrs. Lucrece in the coffee room; Mrs. Lucrece does a version of Miss Tisdale: "Oh, it pains me to think about it. I can't think about it."

Next is Mrs. Baccio, whom I take to lunch. Mrs. Baccio is middle-aged and full-bodied, and, as she informs me angrily, Italian. "Is it because I'm Italian," she asks, "that you think I'll protest? Italians are always the complainers, am I right?"

"Honest, Mrs. B., no," I say.

"You've singled me out," she announces. "I'm demeaned."

We end our lunch early. Maybe Mrs. Baccio is the one who turns me in. Management, in the person of Miss Carstairs, wearing lavender chiffon and piled white hair, calls me into her paneled office. She stares frostily. She emphasizes that the store no longer needs two people in Children's Books. I can see, can't I, that sales are down? And that book supply is down? Because of the war? "You've noticed that, haven't you, Miss Farnham?"

I don't give her my statistical lecture. "No, I haven't noticed it."

She smiles and says I haven't been here long enough to get severance pay.

I'm embarrassed when I call Dorothy. "I was excited about it, Dorothy."

She proves that she's a good social worker by asking, "Are you all right, dear?"

Eleanor says she knew from the beginning I would get fired, because Brentano's is an exploitative enterprise. "But," she adds, "it was worth it, wasn't it? Interesting and instructive, am I right?"

I agree, it was interesting. For the first time since I came East, I think I'm doing some of what I had hoped to do. Late at night, after Ellie's in bed, I talk to my mother about this. I talk to my mother, dead since I was four, on special occasions. She's always responsive. She is older this time; her face has gotten narrow and thoughtful. She congratulates me on Brentano's and says it takes many failed efforts, many many of them, before you succeed.

There are several sequels to my assault on Brentano's. The first is that I become active in Local 27. Dorothy asks me to join both the organizing committee and the publicity committee. I'm happy to serve on these committees, which meet at night in a matey huddle in Dorothy's office. I've got criticisms of Local 27 — that it's too white-collar, too polite, too feminine, not fiery enough. I understand how you deal with this. You get in there and organize. Do it yourself. Well, I'm here and I'm interested. My piston is chugging.

The major sequel to my Brentano's experience is that I meet Lew, the union president. He comes to Washington to testify at a Congressional hearing and asks to see me. "Fine," he says, standing beside me, looking up. (He's shorter than I am.) He invites me to lunch.

Lew is a large-headed Napoleon, fiercely erect in his expensive gray suit. "You've got something," he announces over our crème brulées. "Did you know that?" He reaches across the table to fluff my hair.

I dodge this move, but somehow don't find it completely off-putting. Half an hour's conversation has demonstrated that Lew is brilliant, which excuses a lot. "You've got enthusiasm," he summarizes. "Engagement. Verve."

I want to say, "Oh, big deal," but don't, because the words feel schoolgirlish.

That evening I observe him at the District Council meeting. Words are skimming back and forth; he's in a jurisdictional dispute with the Electrical Workers' Union; he wants to organize the office employees in the new Potomac plant. "Brother Revell," he's asked, "how can you possibly maintain that..?" "Did you even consider that..?" "Did you think at all, Brother Revell, of providing for..?" He listens politely, looking interested, head tucked against his chest, half-smile coming and going. And, finally, he opens up a professional offensive: Slice, slice, a neat row of verbal knifings— you people did this, you people high-handedly did that, you people are not thinking, and the opposition vaporizes. Even dumb, hopeful questions from the audience disappear.

"I don't believe this Lew is such a good idea," Eleanor says. "You've got other boyfriends."

"He's not an idea, Ellie. I'm interested."

Of course, I've had other boyfriends. (We don't call them

lovers, that's too committed and sentimental.) But none of these boyfriends mattered. There was Hungarian Laurence from the restaurant. There was Stephen, a British Army sub-conductor. Lately, I've been seeing Jack, a nice union organizer who writes me letters on wide-lined yellow paper. Lewis packs more punch than these men. He matters quite a bit. I start going to New York on weekends to see him.

He's an amazing conversationalist. I'm not sure I actually like him, but I love those long evenings stretched out on his bed fussing with his Siamese cats and talking about the Theory of Values and the Withering Away of the State. And making love. If love is the right word. Doing a lot of that, actually. Because Lew is an accomplished bed-partner, a repository of tricks and knowledge. None of the Laurences or Jacks understood anything at all about female anatomy.

Lew has a wife who is never around. She seems not to matter.

Suddenly, one evening, he says he wants to hire me as a union organizer. I'm the logical person; Dorothy is moving to Boston.

"And," I ask, "I've been sleeping with you?"

He freezes completely; tension zings visibly through his torso. He turns his face toward me. I'm surprised to see how white it is. "Don't ever say anything like that again. Ever."

This is so much the right answer that I tell him I'll think about it. Of course, I want to be an organizer.

"You don't learn how by reading," he says, when he finds me with a stack of books about organizing campaigns. "Just get in there and do it. Be available. Listen."

So I do, for almost a year. I'm not very good, because enthusiasm and idealism don't suffice. They help only some.

It's an interesting year, though, during which I learn a great many things. I devour a thousand books. (Ellie and I are reading the same ones, and she's the person who keeps score.) I go to four union conventions and learn to jitterbug and to sail and to type. I give a talk about white-collar workers at the Workers' Bookstore. I testify before the Senate Cost of Living Committee on the economic plight of secretaries. I sign up some teachers at the Arthur Murray Dance Studio into the local. I become much better at organizing. "You're nicer," one of the union members tells me. She doesn't know that I've figured out how to listen.

I have lots and lots of Washington friends, and so does Ellie.

She cries sometimes about her husband and goes off to visit him in camp twice, but mostly she is a great companion, funny and supportive. I love her, and she loves her job with the Agricultural Workers' Union. Our apartment is full of people every night.

Lew is part of this life. I see him once a month. He's not my only boyfriend, but he's the real one. I go up to New York on the train; sometimes he comes to Washington. We have good meetings, full of sex and ideas, although the sex is fading some— apparently you need more than technique to keep it going.

And then, near the end of the year, I tell him we're finished. We are through. It's over between us. I'm angry when I say this. It's a sudden pronouncement, without much buildup, but it's heartfelt.

We've been in Lew's Manhattan apartment, a tan-and-green modernist arrangement with a Brancusi at the end of the hall. And the cats, who are affectionate and silly, whom I like. And an absent wife, whom Lew is straightforward about. He doesn't sleep with her. "We're just good friends."

I'm surprised at my irritation that next morning when he departs for a Chicago meeting with the words, "Listen, sweetheart, go over the bed for red hairs, will you? Before you leave?"

I pack up and return to Washington. "I heard you attacked me at the local meeting," he says over the phone. I agree, "I said you were high-handed." Then I make my pronouncement about how we are through. And I wait to be fired. I'm a minor employee, and Lew is a proud man.

But he doesn't fire me. On the contrary, he just says, "Oh," quietly, and adds, "Redhead, it's been O.K., know what I mean?" And two days later, I get a letter from the union secretary, whom I don't know, offering me the job of legislative representative. An important job, a plum. It pays well. We don't have a legislative representative, not yet, but Lew has talked about the possibility. The legislative representative will be the union's Washington person, she'll go up to the Hill and lobby Congressmen and write a Washington newsletter.

"People are endlessly convoluted," I tell Eleanor.

"It's a selfishness-thing," Ellie says. "He has a mental picture of himself, that he's generous and witty and easy. He needs to live up to it."

I say, "Well, he showed True Class on this one. I'll miss him."

So now I am a legislative representative. I have my own office with a glass door and my name on it. I have stationery with my name on it. I attend meetings at the National CIO. I write a newsletter, *Notes From Washington (What Did Your Congressman Say About Your Future?)*. I'm bussed off to local unions in Baltimore or Philadelphia, where I talk about You and the Sales Tax; You and the Cost of Living Increase; Make Your Company Give You Real Vacations. I love my new job. I think I'm pretty good at it. And at the same time, I have moments of wondering whether all this is extra and unnecessary, my pride in it an unwelcome dimension of me.

"Is all this stuff phony?" I ask Ellie, who says firmly, "It's education, Diana."

Everything goes well. Lew stays faithful to our unstated contract; he doesn't try to move back into my life; no checking up in any visible way. My contact in the New York office is with the union secretary, an unobtrusive fat man named Jerome.

And I make a new set of Washington friends—the other legislative representatives. There's Edna, beautifully tan, with a crisp cap of short white curls, who works for the Electrical Workers; Patty, curvaceous and bouncy, from the National CIO office; handsome, rigid Sam, with the Mine, Mill and Smelter Workers. These people spend their free time together and have in-jokes and references implying a secret and superior world. Curiously, I don't feel excluded. They're friendly; they invite me most of the way in; they include Ellie, who is almost one of their set anyway since she's a researcher for the Agricultural Workers. We go to concerts and movies in groups of four or five; Aileen from the CIO Publicity Office is the fifth. She's a nervous, finicky woman who attaches herself to me. She has an itchy habit of rearranging any artifacts at hand—her silverware, her pencils, the pages of a memo. She's the only one of the group whom I don't enormously like.

"All of them are bright like Lew," I tell Ellie. "That shouldn't mean so much to me, but it does."

I tell Patty, the curvaceous, bouncy one, "I have an urge to be better."

"Good sign," she says.

Patty has a real IWW-type labor history. Maybe everyone in the group does; she's the one I talk to about it. Over coffee she tells me about driving a rickety Ford through the California apricot groves, dispensing bottled water and cans of louse spray to the migrant workers. "Wonderful time," she says. "God, Diana. When you feel pulled out of yourself. Made new and different.

"Have another cup of coffee," she invites, and laughs.

So everything in my life is perfect and not quite perfect. I need more. "Not a man," I say to Ellie. "Something for me. What does 'validate yourself' mean, El?"

"Prove to yourself you're worthwhile."

"I *am* worthwhile." I'm walking up and down our hot little, peeling-veneer living room. Ellie is sitting on the couch, listening to me, and occasionally turning a page of *PM*. "El, I do this work and it's O.K. I know it helps people. But what then? I want to look out ahead and see a future. A grand scheme. Something I can feel part of. A framework. Lew used to talk about an architecture. A shape for the future. A guiding plan of ideas for people to live with and set up a new world." I stop for breath and wave an arm. I make an exaggerated, big-eyed, hopeful face. "Something we have instead of God." Both of us laugh, because I'm quoting Hemingway's *The Sun Also Rises,* where Brett spouts about "something instead of God," when she gives up her bullfighter. That was one of our thousand books last year; neither of us liked Hemingway much.

"Listen, I'm talking about Labor, right? Labor, and all we need is for the workers to get together, am I right?"

Eleanor shrugs. "Yeah. Well, of course."

I can see she wants to add a "but," so I go on: "It takes time, natch. I understand. But when all the workers everywhere, when they're united and have a majority in every industry…." I'm running out of steam, my vision has begun to fog.

Ellie says, "So they get together. And organize everybody. Everything, everybody, everywhere. Here, and in California, and in China. Then what? The whole world is in a union. Everybody feels good. Then what?"

Ellie and I have had this discussion before; we've even done it alternating the points of view between us; I know what she's implying: Do the workers continue their good organizing feelings forever, on out into the future, into the universe? Forever? And form a spontaneous government? You need some kind of structure, don't you? "You need a final commitment, don't you? Something to believe in that's much, much bigger than you?" I make these last remarks out loud.

Both of us know we're talking about Communism, although nobody has used the word.

I've been debating with myself for months about joining the Communist Party. No one has approached me, but I know the Party's out there, part of my scene. Surely, most of my new friends are in it. Plus Lew and maybe even Eleanor. She and I talk about everything else, but I'm not going to ask her, flat out, about this. Her husband's in the army, and there are a lot of bigoted people, both in the government and out, who've begun, just in the past few months, talking about the Dangers of Communism. Maybe it's not safe to ask Eleanor.

For me, the word "Communism" remains vivid, scary, heart-stopping, positive. Our world—my whole new, bright, exciting world—is suffused with references to socialism, to the theories of Marxism, to quotes from Leninist writings, to bits of Marxist history. It's shot through with these the way an Impressionist picture is suffused with light. The threads show up everywhere, in novels, plays, songs, newspapers, in *PM,* even in the staid *Washington Post.*

"I've got to be in the forefront," I tell Ellie. I've stopped pacing the living room. I'm standing still now and scowling at the frayed furniture.

The remark sounds adolescent, but it's a true part of me. I want to go all the way.

"I've fiddled around long enough," I say. Eleanor just looks at me.

Then Ellie decides to leave Washington—her husband's being sent overseas. She'll work in a war plant, maybe in California; she'll be one of those poster-women wearing a kerchief and gesturing with a wrench. "Jesus, I'm going to miss you," I tell her.

Joining the Communist Party becomes confused in time and space with Ellie's planned departure from Washington.

I speak to Patty. "I've been feeling I need to go further. In my beliefs. In my commitment. If you know what I mean."

We're in the hall of the House Office Building. She looks at me speculatively.

And I wait for something to happen.

For a while, nothing does. Then there's a clot of invitations from people I hardly know and don't care about, a curious procession of inquiring faces. The faces say little and seem to be feeling me out. Patty's secretary arrives. Also, the manager of the Workers' Bookstore. And a researcher for the *CIO News*. They offer me lunch and afternoon daiquiris and watch me from under their eyebrows.

Until I hear from Aileen, the only legislative representative I don't truly love. She takes me to the Shoreham Hotel, aims an accusing stare at me over a shrimp cocktail. She fiddles with her silverware, reorganizes it to get all the edges lined up, and says, "Diana, some people are afraid of the ultimate commitment. But I'm sure you aren't, are you?"

I hadn't thought of Aileen as my bridge into a new life. I try to imagine an Aileen-Marxist world—conformed, mechanical, boundaries measured, all the pages with perfect margins. Yes, there can be something rigid about Marxism.

This whole drama has started to make me nervous. It reminds me of my efforts at Brentano's.

I don't say, "You people are doing this ineptly." I also don't say, "Back off, will you, please?" I look around the Shoreham's peaceful outdoor terrace and tell her, "Aileen, I thought I knew what I wanted. Now I'm not sure."

The next day I call Lew and recount the story to date, more or less circumstantially. I say, "Lew, get them to slope off for a while, will you?" I don't question for a minute that Lew will understand and be able to do this.

And two weeks pass with no further recruiting.

And eventually there's another overture. The woman who arrives at my office is a stranger—spare, shy, and New England, a washed-out blonde wearing a rumpled summer suit. She peeks in, shuts the door behind her, sits down and looks at me hopefully. "I'm Elizabeth. I'm Party secretary."

And she invites me to an evening event at her apartment. "Just some friends," she says, blushing.

Elizabeth reminds me of Dorothy in Local 27. I like these sincere, tongue-tied women.

There is no one I know at the party in her pleasant third-floor apartment. I survey the ten or so people circulating among the green wicker sofas and standing, glasses in hands, on the worn oriental rugs. They're inoffensive Washington types who smile nearsightedly in my direction, but avoid eye contact. Maybe I've seen a couple of them at the Workers' Bookstore or at a District Council meeting.

Finally, a weedy fellow separates off and corrals me onto a couch beside him. "Drink?" he asks. "Red wine? White wine? Hi, I'm Roy. I'm district organizer."

I like the fact that he and Elizabeth are officials, with titles. At least I know whom I'm dealing with. And I think I understand Roy, who's not like the others. He's worn-looking, with a narrow red face and protruding teeth. He could work in a gas station or a restaurant kitchen; probably he actually comes out of a factory. Roy is the real thing. A genuine worker.

He starts talking in a flat Southern voice. "We really need you. You need us. We been watching you a while. You're really good." He's not saying the right things. But how could he know that I'd react best to a grand vision, to the story of a poetic or religious future or of a life larger than my own?

"Another drink?" he asks. His approach isn't sexual; I'm grateful for this.

I survey the room. Roy senses a pause. "All these people," he gestures, "they've exposed themselves to you."

I want to be cross. "I wouldn't know them again if I tripped over them." A roomful of strangers. Some well-dressed and some Bohemian-draggled. The Party is playing its cards close to the chest.

But I think some more and look again and, yes, these sincere, strained people were ready to come forward and let me stare at them. His point is serious.

And, honestly, I am ready.

I truly want to make a definitive gesture. "O.K.," I say.

Joining the Party involves signing a card, just the way legend tells us. Roy produces it out of a pocket, a blue-and-white card with "Communist Party of the U.S.A." across the top. And I sign it right then, sitting on the edge of Elizabeth's rattan sofa, shaking her balky Sheaffer Lifetime pen to get it to write.

I become a card-carrying Communist. □

Diana O'Hehir lives in Tiburon. She taught at Mills College for 34 years and is the author of four books of poetry and two novels. This is an excerpt from a memoir-in-progress; her first mystery novel will be published later this year by Berkeley/Penguin. E-mail: dfoh@sbcglobal.net

Primero de Mayo

Yvonne Martínez

*W*hen I was six, I stood up to my mother's boss. I remember going with her to pick up her check at the Salt Lake City laundry where she worked. She and the boss exchanged some borderline unfriendly boss/worker banter that I finished, unseen, from the other side of the counter.

The boss paused, peered over the counter, and said to my mother, "You'd better smack that girl, Margaret."

My mother looked past him, beamed defiantly, took her check, put it in her purse, grabbed my little hand, and strode off.

I had said what she couldn't say. It was one of the few times I remember not being yelled at for saying something.

As a Latina in Salt Lake City, where the term *Mexican* was twisted into a slur, my second-generation-desperate-to-belong mother so identified with the assimilationism of her generation that I was constantly admonished to comb my hair straight-unless she was trying to shape it into Shirley Temple ringlets-and scrub my knees and elbows so they would wouldn't be so dirty. She meant dark.

The Mormons made sure that we all knew our place. Even my half-Mexican Jack Mormon cousins had a clearer way to heaven than we did. They belonged, kind of.

I didn't. When my class went on a field trip to see the Mormon Tabernacle Choir, I was separated from the rest and kept outside, because there were sanctuaries in the Temple that I was not allowed to enter. Then there was the show-and-tell day my pink plastic rosary was broken and thrown on the floor under my desk. I picked up the pieces, gasped, held the tiny pink beads tightly in my hand, and looked around. There was no one to tell.

Because of my out-of-wedlock birth, my mother had been forced to "marry any man who would have you." But, she was warned, "whatever you do, don't marry one of those from *el otro lado* (the other side)." My very old-school stepfather was in this country illegally as a lapsed *Bracero*. Like the others, he had been brought here under contract with very limited legal protection. He lived in what can only be described as apartheid-like conditions. The Braceros were single men, not allowed in the towns, confined to the labor camps. My stepfather wanted his part of El Norte, too, but he was not suited to farm labor. Leaving that behind, he ended up in Salt Lake, where he met my mother, who was a waitress in the family restaurant, which also served as a dance hall and bar, with a rooming house upstairs.

My mother's family was only a generation away from the peasant squalor and misery that led to a revolution, but they kept their distance from the *rasquatchies*, the greenhorns. Excluded from doing business with whites, these families held on by supplying the needs, legal or otherwise, of the inhabitants of the Mexican quarter. The greenhorns had been the chief source of labor for the Utah & Idaho Sugar Company, and, with Puerto Ricans, had worked the local mines and smelters. My stepfather had fled revolutionary violence in Cueramaro, Guanajuato, to work in the mines of Ely, Nevada, and the beet fields of the Uintah Valley.

When I was eight, we moved to South Central Los Angeles where my stepfather got a union job on a repair gang with the Santa Fe Railroad. My mother found work as a paper folder in an envelope factory.

Before we left Salt Lake, we had a visit from my Grandma Mary. She had organized "hospitality" workers during the forties and fifties. Her father, my great-grandfather, was the Millard Bandit, gunned down in 1922, by a posse of 90 men in Delta, Utah.

"You'd better come and get Mary," they told my mother.

"She's shut Frank's down again."

Before my mother could dispatch my stepfather to the tavern, a taxi pulled up. Grandma twirled up the steps like a top spitting sparks.

"Those bitches were giving it away for drinks again. Pinche Frank. I work there because I need to work. I'll shut that hijo de la chingada down again, if he brings those chippies back. Trying to cut us out, cabron. I cost him more today than he'll make in a week." Coat off, flopped in a chair by the fire, she commanded, "Get me a beer." Looking around at our stunned faces, she demanded, "What are you all looking at? This ain't no bingo game."

We stood back. Calmer now, she drew me close to her. Hands on my face, she said, "Mija, some day you'll learn the difference between a whore and a working woman."

When we moved to Los Angeles, it had been decided that it was my turn to redeem the women in the family. I was to be sent to Catholic school. There is a picture I love of my grandmother in her swing coat, standing next to me in my first-communion dress. From the priests and nuns — or *numbs*, as I sometimes mistyped them — I learned that my life had value in community and in the service of justice.

The first time I stood up to my own boss, I was 17 and worked in a downtown L.A. theater after school and on weekends. I heard that the cashiers at the West Side Theaters had a union and better pay. Most of the Black and Latino kids I worked with worked to supplement their families' income and in some cases supported themselves entirely on what they earned. We needed a union, but didn't know how to get one.

My job as a cashier was considered a premium job. My boss had a habit of promoting the young women he liked into the box office and then into the bedroom. When I resisted his attentions — and questioned his money-handling methods — I was fired. I had no other means of support and was living on my own.

I left that day and walked to every downtown theater in the chain, determined to find work. Word had already gotten out that I had had a fight with the boss, and no one would consider hiring me. Finally, a manager at the Warren's Theater agreed to use me as a sub, temporarily, until he could work things out with my boss.

He knew I was a good cashier.

Ultimately, in order to get my job back, I had to apologize to the boss who'd fired me. It was humiliating, but I got my job back and no unwelcome demands were made of me after that. The fact that I was almost blacklisted was a hard lesson.

The feeling of being cast out, scorned, punished, and blamed by my boss-perpetrator was no worse than the humiliation of being forced to go back. That feeling never left me, although I didn't realize how much I had been changed until much later. It was as if something inside of me had been readjusted, recalibrated, and retimed. For the moment, even though I'd lost my job and didn't know how I would pay my rent, I was free. I had ripped through fear.

Standing up to a boss is one of the most difficult and life-changing things people are ever called upon to do. Standing up to a boss can be a metaphor for standing up to any kind of power. Sometimes it's as simple as telling the truth to power.

"All you have to do is notice and say," I heard Portland First Unitarian minister Marilyn Sewell say one Sunday. In that sermon I also understood her to say that if sin is a denial of self, then ours as women, workers, people of color, the disadvantaged, oppressed, and downtrodden of any kind, is a sin of omission, a sin of silence and self-effacement. Ours, she explained, is not a sin of arrogance or grandiosity. We sin by not noticing and saying, by not accepting what we know and by not taking action, even if it only means speaking up. Just putting the words into the air without trying to predetermine the outcome can be a revolutionary act.

At each new contract negotiation, I tell the newly organized workers that the process is not about how fancy I dance at the table or how well I articulate their concerns, it's about power, plain and simple. It's about what the boss believes will happen if power is abused. It's about putting your bright yellow contract on top of your computer terminal so the boss can see an ocean of contracts as he walks the gangplank above. It's about taking a seat on the bus when you're tired after a long day's work and being prepared to be arrested. It's about walking out of a strawberry field when you know that you may not have earned enough to buy something to eat. It's about walking off the ward knowing that it's ultimately the hospital that is forcing you to leave your patients behind.

In 1994, I was the recipient of the AFSCME Jerry Wurf Memorial scholarship to attend the Harvard Trade Union Program. Established in 1942, the program is a once-a-year ten-week residential program. At this training, my counterpart in the Union of South African Mineworkers commented to me upon learning the specifics of our labor legal system, "In the worst of apartheid we had better labor laws than you do here." Another classmate, a trade-union activist and separatist from Quebec declared, "In the U.S., you don't need a law, you need a revolution!"

My Australian and Japanese classmates were shocked to learn about our bloody labor history—the massacres, jailings, beatings, trumped-up charges, deportations, etc.

Our labor history is violent. Our labor movement has been racist as well as sexist. Thurgood Marshall began his legal career with a lawsuit against a railway union, because the union refused to represent a black member. It was common in those days for unions to have Jim Crow union halls, separate seniority lists, and exclusionary charters. Samuel Gompers refused to extend a charter to include Asian workers who had successfully fought alongside Latino workers in an early California citrus-industry battle. The Latinos refused the charter. Marshall won his case, and now unions have a duty to represent members regardless of race, sex, or creed.

Until the recent election of Linda Chavez-Thompson to the national AFL-CIO executive board, union leadership reflected the diminishing ranks of union membership: it was moribund, white, and male. At its height, 50 years ago, union membership nationwide was 35%. Now, with only 15% of the workforce organized into unions and with tremendous demographic changes in the work place, labor has had to wake up. And it has.

Linda was in Oregon recently to campaign for a local politician. "I'm here," she said, "because it makes no sense to send two men [referring to newly elected national president John Sweeny, of the Service Employees, and vice-president Richard Trumpka, of the Mine Workers] to do a job one woman can do." We hooted, shouted, clapped, and stamped our feet. We had finally arrived. We've been doing double- and triple-shifts for years, and Linda knew what that meant.

This was Yvonne Martínez' first time in print. She currently represents the Communications Workers of America at UC-Santa Cruz. She notes: "This piece is dedicated to the memory of Jimmy Bryan. I welcome and encourage testimonies from union workers. This essay was written not to punish, but to change, and out of my deep love for working people. I have changed some of the names, but none of the facts. Special thanks to The Flight of the Mind Women's Writing Workshop for unwavering support and sustenance." E-mail: thsstrhd@hypersurf.com

DIGGING

Toni Mirosevich

> *We either forgive each other*
> *who we really are*
> *or not.*
>
> Ralph Angel

My mother, Vera, and I sit on the sofa in her tiny apartment in Anacortes, my city of origin, a small town in the great Northwest. Well, maybe less great than it used to be, as are these digs she's lived in since my father died on the fishing boat and took the big house and the dreams with him. We are watching the evening news, an activity we slip into happily, something to fill the air, even though it's the first night I'm back home for a visit and we shouldn't have already run out of things to say. There is fighting in Sarajevo, in the former Yugoslavia, our land of origin. Her parents grew up on the Dalmatian coast and left family and friends behind to give it a shot in America. So, it's a country that's once removed for her, twice removed for me. It's all in how you measure distance.

On the TV, an anchorman leads a camera crew down into a church basement that currently serves as a makeshift hospital. The worn faces peering back at the camera look eerily familiar. One old man on a cot, around 70, about the same age as Vera, has a bandage sashed around his chest. He is the spitting image of my Uncle Ivo—the same nose, ears that radar out of his head. But Ivo got out long ago, made it to the new country during Tito's reign, when you could still leave. My mother always said Ivo got out while the getting was good.

The reporter addresses the people in Serbo-Croatian and, before the BBC voice dubs over, for a few small seconds, I can comprehend a word here and there. I can make out *ništa*, which I know means "nothing." Maybe the question the reporter is asking is, what do you have now? Ništa. What do you hope for? Ništa. What does the future hold? Ništa, ništa.

The English translation kicks in. A man who lost his left arm in a bomb blast tells the reporter that the rest of his family is gone. When asked if he would now take refuge in another country, he says, "Why would I want to leave everything I have?" He spreads his good right arm wide to the horizon, as if to make a comment, to underscore. The camera pans a countryside pockmarked with mortared homes and bombed-out farmhouses. His eyes are shining with some weird belief, like a stage mother, like the parent who looks at her homely female child, lumpy and lopsided, and sees a future beauty queen.

The picture shifts to a hillside with columns of gray smoke. There's a popping sound of gunfire, then one blast, then another. My mother looks up from her knitting—green-and-blue checkerboard slippers for the homeless shelter—shakes her head back and forth, then returns to her task.

Usually, we don't touch it, this present past, how what's happening right now in the old country figures into our current lives. She never volunteers an opinion about the war—as if it's too revealing, as if it's like telling a stranger some shameful intimacy. Maybe if you want to leave the past—the accent, the foreignness— and stamp your children American, then it's best to never refer back. You can only address the old country in nostalgic moments, at the end of the meal, when you're with your kind, when gush is forgiven.

There they are, my kind, on the coffee table, assembled in a jumble of picture frames. Not an inch of table top shows. The frames huddle together to form a little community, as if the people inside are seeking warmth and union. There are pictures of my married sister, who lives close by, close enough to be there for Vera's health emergencies; pictures of my unmarried cousin Mate; Teta Barbara; a family picnic in a park; Uncle Ivo in front of the Croatian Hall. Looking at them, I remember how as a teenager all I wanted was to be anything but Slav. The blonde seventh-grade girl I had a crush on was named Marcia Darrington. I used to say her name over and over just to hear the ring of it. The name suggested neat edges, the upper crust, the final suffix, *ton*, an English word no less. In our community, everyone was an *ich*. There were other differences, too: Getting an education was making it to secondary school; climbing up and out was the move from blue to white collar, the chance to stay clean through a full day of work.

Since I moved—to get away, to get educated, to live a modern life—my visits home have always been brief, quick fly-bys to maintain connection, but never long enough to ruffle. We only have so many stories to tell.

When I return home to visit, with my multi-syllabic words, my educated gift of gab, it's as if I've forgotten our common language, the way the voices at the dinner table rise and fall, the song of it. Maybe, too, I've forgotten a basic Slav know-how: how to fillet a fish, how to tell a good person from a jerk, how to touch people when you talk to them—on the arm, the leg—all the broad gestures, the broader emotions, how to sing.

How can I make my mother hear this, the dissonance I hear when I come home again, the test of each return? How can she ever understand who I've become?

There's a picture of a Serb soldier in a trench, a cocky angle to his helmet, a guy's guy. Rambovich. It's all too dramatic not to note, not to take a stab at.

"What did your mother tell you about the old country?"

This is how I always start. If the question registers, she's not letting on, she just continues to knit, to grow green-and-blue rows. I am used to her stonewalling about the past—the painful re-entry into memory, poverty, troubles. I have tried to analyze it: Is it that making a link to the past might bring the conditions back? Is it that to remember strife is to welcome it again through the back door?

"She must have told you something."

"She didn't say. She was always pining."

Pining is such an old-fashioned word. A word that conjures images of fainting couches and widow's walks and shawls thrown around shoulders against the chill of the sea. Someone cold and left. The word should silence me; it should be answer enough.

"She had to mention something else." I'll risk it. It's true we're of a different age. I revel in this type of excavation. In therapy, I have always mined any nubbin of a mystery, have processed out what's hidden. But even I know what thin soup a word like "process" is, know that rumination is the luxury of those of us who work with our heads, not our hands. What equivalent did she and her mother have? It's hard to imagine Lina, my Nana, uneducated and silent, dressed in eternal black after her husband died, sitting beside Vera on the couch and having a heart-to-heart.

The slippers are taking form. I try to picture the Bosnian man on TV with checkerboard feet. Better to knit a rifle sling, or an ammo belt, or a grenade. Think big: Knit a bomb shelter.

She fingers the channel changer, itching to switch to *Jeopardy,* though there's a good 15 minutes of the news left. It's not like I haven't seen this maneuver before—my mother's personal flak jacket. When I protest, she says, "I watch it to keep my mind sharp." I hate having to endure a half hour of Alex Trebek, the whistles and bells when a contestant picks the Daily Double, all those questions about art and politics and history. I want to shake my mother awake, tell her: Here's the real news, here's history. Look at these people, here's the real Double Fucking Jeopardy.

She gives me a look with a sting to it, the same look she gave me the time I called my boss an asshole. She tried to be sympathetic— any family member against the world—but then volleyed her hardest stone. "Life's not fair," the usual platitude, then, peering at me out of the corner of her eye, "What made you so hard?"

The picture on the news is of a kid with no legs. Let's talk fair. How much time is left? What happens if everyone goes, what happens if they all drain out? Already the Serbs have bombed Vukovar, Dubrovnik. With every bomb, a piece disappears. And the relatives who made it here, who hold it all—memory, history— like money stuffed in a coffee can and buried in the backyard? Her older brothers and sisters have gone, to cancer, to diabetes. If I tally up my mother's current list of ailments, the odds aren't good: two small heart attacks in the past year, a bleeding ulcer, phlebitis in her left leg. Events I'm never present for, though she took care of her mother forever, the old-country way. Each Slav house always had a spare room for the mother's eventual return, prepared with the embroidered pillow cases, the easy chair, ready for when the time came. A room I'm reminded I lack, when Vera comes to visit my rent-controlled studio.

It's true, I'm not always forthcoming when she wants to be let in on my life. There's the let's-ignore-it gay thing, the lack-of-direction-when-are-you-going-to-find-a-good-job thing, the how-can-you-live-in-such-a-hovel thing. I'd rather tell her fiction and obliterate the facts. Or at least hide them. Maybe it's in both of us, when it all gets too much. The desire to rewrite a chapter, or abolish it altogether.

There's a commercial for a two-ton pickup. Then one for antacid. Then one for a Friday evening program schedule called TGIF. Maybe I'll suggest we have a drink, an early cocktail. Sometimes that eases things.

My mother gives a little cough, some signal, shifts in the chair, then clears her throat. Then whispers something, barely audible:

"Digging."

"What?"

"Digging. Nana said there was digging."

I leap on it. "What? Potatoes, troughs?"

She picks up a new skein of yarn, magenta-colored, and starts to rewind it.

"Just digging."

It's up to me to imagine. I've seen pictures of the old country, the land mass edged by the Adriatic, too blue to look real. The rocky soil of the islands must have made it particularly rough, the shovel hitting stone. Or bone? What could you grow in such unforgiving earth?

"Anything else besides digging?"

The news anchor is reporting the Dow's hit a new high. The needles hit each other a little harder, like heels on a waxed floor, rat-a-tat-tat. Her face has a granite look t it. For a minute, Vera looks like her mother, as immovable. How will she forgive me my prodding; how can I forgive her silence? It's what we've been doing all our lives, our ritual. But there's something else. On down the road, how will we get back at each other? Only when she was in a rest home was her mother forced to wear the garish-colored bedrobes her daughters brought her. A payback for all those years of black.

"Fiestas. They worked hard, and then they had fiestas."

Hard to imagine Nana, her black skirt swirling, as the tamburitzas played.

I remember Nana here in this country, moving around in her vest-pocket garden. She never had much to say to me. But once, when I visited, in the midst of some deep high-school funk about the meaning of existence, she took me aside. We were both dressed in black—my nihilistic uniform, her widow's dress. We edged around the plants in silence. Here and there she'd bend over, scrape the dirt around the base of the greens or rake at the soil. Then, out of nowhere, first in Slav, then in broken English, a kind of all-purpose advice: "*Malo po malo.*" Little by little. She said, "Go out and put your fingers in the dirt, under a plant. Move the dirt around. Malo po malo. Malo po malo. Dig, a little."

The news shifts to Rwanda, another makeshift hospital, another language. The news anchor gets on and makes his final commentary. "With time running out—in Bosnia, in Rwanda, in Azerbaijan—what will be our official answer to these hotspots around the globe? In the face of rising tensions, what will we ultimately be called upon to do?"

It doesn't seem to register. My mother gets up from the chair, puts her knitting down, gets ready to make dinner. But, as she moves past the TV, she gives a little parting shot. "Forgive faster," she says, then flicks the TV off. □

Toni Mirosevich is an associate professor of creative writing at San Francisco State. Her new collection of poetry, Queer Street, *will be published this spring by WordTech Communictions, Cincinnati, OH. E-mail: tonimiro@sfsu.edu*

TOO BUTCH FOR ME

Judith Barrington

A few years ago at a writing workshop for women that I direct, there was a mystery-writing class with twelve students. When the group was faced with sharing their work at one of the evening readings, they were at a loss as to how they could all read their longish, plot-driven pieces.

After some discussion they decided to appear as a whole group, each member of which would read the first and the last sentence of her mystery. The only question was how to arrange themselves in a line in front of the audience. Their line-up became a mystery in itself, with the rest of us required to guess the criteria they had used to create the line. As they stood there under the spotlights, the audience was indeed mystified: the twelve women looked as if they were in some logical order, yet what was it?

It wasn't age. It wasn't height. It had nothing to do with clothes. We just couldn't guess. Finally, one of the writers explained that they had used the Kinsey Scale: they stood in a line going from the most heterosexual to the most homosexual. When they had decided on this, they told us, everyone in the group had immediately known where she belonged.

Ask a group of lesbians or gay men to line up with the most butch at one end and the femmiest at the other, and I bet they'll have no trouble knowing exactly where each person belongs. But ask them how they know, and they will simply look confused.

I am what is sometimes called a "soft butch." I know some of my friends, particularly the straight ones, will argue with me about this—I'm sweet and not very loud and I don't trail chains, though I do sometimes wear a tie—but the fact is I've been butch for as long as I can remember.

Trying to define butch and femme is as hard as trying to explain what makes me wake up at dawn and droop at 10 p.m., while my partner will happily sleep till mid-morning and socialize into the wee small hours; or why I adore clear blue skies and hot sun, while my friend Phyllis comes alive under a blanket of cloud, deep in a drippy forest. Some differences just are.

In the early seventies, when I was coming out, one recognizable style belonged to the lifelong lesbian, whose experience had told her she had no choice—that it was probably genetic or at least in some way biological. Then there was the political lesbian, who thought that her beliefs, not her genes, led the way to lesbianism. Back then, a political butch was likely to develop a different look than an old butch bar-dyke, just as a lesbian-feminist femme looked and acted different from a femme who had lived through the fifties in the closet.

Now in the nineties, activist lesbians are more inclined to wear tailored pants and good jackets if butch, and drapey skirts and good jackets if femme, although black leather and body piercing are the choice of many younger radical lesbians, among whom the butches look fairly similar to the femmes. Closet lesbians continue to look depressingly like they looked 20 years ago, the butches in particular displaying that familiar awkwardness in their camouflage outfits, but modern lipstick lesbians, whose numbers include a good many butches, span the political spectrum and are often brazenly out.

Even harder to recognize as butch or femme is the woman who feels genuinely bisexual, but tends to like women better than men, or the politically motivated bisexual attracted to both butches and femmes of both sexes. Still, I stubbornly cling to my belief that on some level we all know who falls where on the spectrum, at least within our own cultures and classes.

Just to add to the confusion, the new category of "transgendered women" appeared on the scene in the nineties. These women, butch as they come, take normal butch defiance to new heights, although this sometimes looks distressingly similar to the old Man-Trapped-In-A-Woman's-Body theory that I first encountered in *The Well of Loneliness*.

What is new, though, is a blessed absence of the shame that permeated the pages of that book, an acceptance that allows transgendered women to embrace openly not only the butch role, but the male one, too, in many respects. I can't help feeling, however, that there's something about the transgender identity that defies feminism—something that indulges and perpetuates the old and often damaging equation of butch with male. The feminism that saved me was also a feminism that claimed a female world large enough to embrace even the most butch among us.

If you want to know about the difficulties of being butch, go with me into a public restroom. Because of the butch-equals-masculine conspiracy, women have been known to scream or gasp as I entered. And believe me, I really don't look like a man. More commonly, a woman busily arranging her hair or her face at the mirror will catch sight of me with an almost imperceptible shudder of fear. As she checks me out more carefully, she will smile to make up for her moment of horror, or become extra friendly, chattering about the weather as if to say, "See, I knew you were a woman all along."

In the old days, I used to blush and feel somehow wrong when this happened. I took to creeping into restrooms with extreme caution and smiling with an almost pleading expression as I did so. I became so afraid of public restrooms, especially in places where women would be dressed up to the nines (the opera, for example), that I would deliberately wear something that flashed a bit of cleavage, and would enter the room humming at a high pitch, broadcasting ahead of me my right to be there. Anything to avoid those startled stares.

Not long ago in Mexico, I went to the señoras room after dinner at a fancy restaurant in Puerto Vallarta. I was with a large group of women, all of us celebrating the end of a class I had taught there, but I was the only one who headed for the restroom before we left. I barely noticed a couple of waiters standing near the door and eyeing me as I entered.

Sitting quietly in my stall, whose door ended a good two feet above the tiled floor, I saw a pair of very masculine-looking feet enter the restroom and stand still for a couple of minutes before leaving. I took my time and emerged five minutes later, almost crashing into the restaurant manager and a policeman, both of whom were clearly about to enter the women's room, of which I had been the only occupant. They stared wildly at me, the manager opening his mouth with a determined look, but the policeman laid a cautionary hand on his arm. Then they both looked me up and down again, their eyes lingering on my breasts. Broadcasting some hostility, they stood aside as I passed between them. Not a word was exchanged between us, but I knew that I had almost been arrested for my choice of bathroom.

Some people ascribe these mistakes to the fact that I am six feet tall, but I know this is something that happens to butch women who are small, thin, fat, friendly, or reserved. When I walk into a store and the clerk, without looking up, says, "Can I help you, sir?" I know it is not simply my height that has spoken for me, but some other nebulous aspect of how I take up the space in front of the counter, what I expect from the clerk—or maybe it's a simple matter of biology. Scent accounts for a lot of things we don't understand; why not this, too?

Sometimes, when addressed this way, I look around me with exaggerated surprise, as if searching for the "sir" in question. At other times, I say testily, "I'm not 'sir,' if you don't mind." But still I have to steel myself to go into bathrooms. One day, I swear I'll get right up in the face of the poor woman who has just emitted a muffled squeak or a gasp. Icily I will stare into her eyes as I hiss: "What's the matter? Never seen a self-respecting butch dyke before?"

I remember a conversation I had with a group of lesbian friends in which it became apparent that the butches all had some kind of fantasy involving rescue. Mine was centered on the woman who lived next door to us when I was ten or so. Her name was Molly, and she was the young, single mother of a daughter about my age. I thought she was the most beautiful woman in the world and would hold imaginary conversations with her all the way home from school.

Every day, I rescued her from some dire situation. She would be tied with a thick rope to a tree while a rabid dog or a wolf with phenomenal fangs snapped at her. I would wrestle the animal to the ground, send it on its way, and untie Molly. Or she would be swimming in a rough sea, almost drowning from exhaustion, when I pounded my way out through the breakers, took charge, and towed her to shore.

Then there's the whole notion of "passing." You have only to read the personal ads in any gay male publication to see how important it is to many of them that they find a partner who looks straight: they advertise for "straight looking," or "masculine," or "straight acting," partners. Lesbians, too, find it necessary to pass, some of them almost all the time, others just now and again for convenience. But it's the butch women (and the femme men) who have the hardest time passing, since men are expected to be butch and women femme. Even when a butch woman thinks she's passing, it's often painfully obvious what she is doing. Of course, making the effort to pass is sometimes enough to ward off the hostility that might be directed at her if she simply embraced her butchness.

An interesting addendum to the question of passing concerns the recent lesbian baby-boom. I can't say I've conducted a serious survey, but it looks to me as if far more of the artificially-inseminated mothers are butch than femme. My friend Sheila, a butch with no interest whatsoever in babies, claims that this is a manifestation of butch oppression—a way for a butch to pass, since motherhood is undeniably central to the role that real (that is, heterosexual, femme) women are supposed to play. This is not to say that butch mothers look at their desire for a child simply as a ticket to acceptability. But an unconscious component of that desire, that search for the highly-touted "fulfillment" of childbearing, might be an urge towards the easier life of automatic inclusion.

Since I've never had a baby, I can't say for sure if it works, but I do know that other traditional, "real woman" activities can buy a few moments of unfamiliar acceptance. One of the most extreme acts of passing a lesbian can perform, for example, is getting married to a man. I got married at 26, because I believed it was what I wanted (self-delusion is an integral part of wanting to pass). I had already had four major sexual relationships with women, although I had never said the word "lesbian" and none of my lovers had considered herself a lesbian. All were older than me, and all encouraged me to get married.

It was a bright, windy morning in June and I downed several large brandies before leaving for the church in my long white wedding dress to marry a man I loved like a brother. Three of my former lovers—all women—sat near the front of the church, crying. My hair got ruined by the wind as the wedding group stood outside for the photographers. The reception was an elegant affair at a fancy hotel with tables set outside in the rose garden.

It's hard to describe my own state of mind. The closest I can get is to compare it to a dream. It had internal logic, but didn't fit into the rest of my life in any recognizable way. I felt as if I were reciting my lines, not only when I repeated the wedding vows in front of the congregation, but also when I chatted with the guests, describing plans for a house and a routine in which my new husband and I would both work. It sounded reasonable. But I never once believed in it, and, eleven months later, when my husband yelled at me through the locked bathroom door that I'd better get out of there and cook his breakfast, it felt only right to seize the moment, pack a bag, and walk out.

In the weeks prior to the wedding, though, I found myself attracting a kind of public approval I didn't even know existed. When I went to pick up the necessary forms, I was beamed at, asked in detail about my fiancé, and patted on the hand, the head (when sitting), and other parts of my body, by a variety of clerks and secretaries, all of whom seemed delighted that I was becoming one of them. This was an entirely new experience and would have been enjoyable had I not been troubled by an insistent sense of fraud. I had never been "one of the girls"—not in offices where my co-workers exchanged boyfriend stories, and not even with my closer friends who had dealt with my reticence by believing that I was seeing a married man. Now, all of a sudden, I was a normal female, not at all reticent—babbling, in fact, about my approaching wedding to all these friendly strangers, but I couldn't really enjoy it. On some level, I knew I was faking it.

Some of my friends and acquaintances display extreme irritation with the whole notion of butch and femme. While several have eagerly discussed their rescue fantasies or recounted instances of passing or failing to pass, others think it is just plain backward to claim a role—they say everyone can be anything she or he wants. And in a way that seems to be becoming truer: young lesbians are more androgynous, less obviously butch and femme than ten or twenty years ago. Yet I cannot fail to notice, somewhere underneath that androgynous surface, the continuing background hum of a difference that is not gender, not sexual preference, but simply the latest style of butch and femme. All-black clothes, distressed jeans, nose rings, chains, leather, or flowing skirts—two young lesbians can adopt exactly the same style and yet one will be recognizably femme while the other is a new version of baby butch. Whatever these categories consist of, they endure.

I'm just leaving the I.C.A. Gallery in London where I've discovered the extraordinary photographic self-portraits of Claude Cahun. Tiny, postcard-sized self-portraits show Cahun with a shaved head, her facial bones like a metal sculpture, or Cahun with hair, softly gazing off to one side. Then again, she's formally dressed in a tie, looking like the Paris-thirties butch that she was.

On the wide path that flanks Horse Guards Parade, I run into Jaimie, who is heading into the exhibit. The last time I saw her was the morning after we slept together, 21 years ago. She's no longer wearing her black bowler, and her face looks more like a bloodhound's than it used to.

"Is it really you?" she says.

"No," I say, "I'm someone else."

She smiles and for a moment I remember why I liked her before I let her disappear without trace. It's the wicked glance that follows the smile, turning us into conspirators in some irresistible plot that she's cooking up. I am about to melt into effusive apologies for my silence over the years, forgetting that she, too, has remained silent, but then I remember I am still displeased with her.

When she suggests tea, we go back into the gallery and sit at an iron table under an Andy Warhol print. I study the new lines around her jawline as she tells me that she's still sculpting and teaching every winter in Key West, where she has, for most of her adult life, spent at least half the year, with many summers in London. She had seemed so cosmopolitan when I joined the writing group; so committed to her art; so startlingly different from anyone I had known until then with her Jewish New York accent and her willingness to say shockingly true things about herself. At least I thought they were true at the time. Later, I wondered if it was just an act.

I'm starting to notice that she hasn't asked anything at all about my life, noting sourly that she's just as self-centered as ever, when she leans forward with that intense look I remember. Back then she would often couch the most acerbic comment in the form of an innocent question.

"So what was wrong that night?" she asks.

"You were too butch for me," I say. I've had a long time to come to this conclusion.

"Too butch!" she says, exasperated. "Too butch! I don't believe in that stuff. I was just being me."

"Well," I say, "the you that you were being was too butch, whether you believe in it or not. And it wasn't just that night. It wasn't about the sex."

"Oh," she says, a little deflated. She wanted it to be about sex.

"Tell me what you mean, then. What do you mean by butch?"

So I remind her of how the writing group met week after week and how she told us that her literary masterpiece was her journal, which she'd kept for 30 years, and which was stored on microfilm in three different countries in case of fire or flood.

She interrupts me. "Is that supposed to be a butch characteristic?" she demands.

"I don't know," I say, "but when you asked me to read portions of the journal, and when you told me I had to come to your house to do it and that you would be somewhere else, and when I sat at your kitchen table reading page after page of your fantasies about me—how you brought my favorite cheese to group meetings and how you couldn't get out of bed for a day when I forgot to call you—then I think you were being butch."

We order more tea while she digests this. I don't try to elaborate on why her past manipulations seem butch to me now. I can't. But I tell her how angry I felt there in her kitchen and how weird it had been to end up in her bed a week later.

"I felt as if I'd been flattened by a bulldozer," I say finally.

"Oh," she says. "Was I too pushy?"

I get up to leave, suddenly remembering how much I had longed to comfort her as I read those pages of loneliness and yearning, how responsible and guilty I had felt when I slammed the journal shut and ran out into Tottenham High Street, my rescue fantasies clashing head-on with my need to feel in charge of the seduction. I lean over and kiss her cheek.

"Don't worry about it," I say.

As I walk down the wide avenue towards Buckingham Palace, I reluctantly recall the distinct thrill I had felt as I drove to her flat that night, wanting to respond gracefully to her hot pursuit, and knowing that my task was merely to acquiesce. But acquiescence lay too far outside the role that was not merely a name, but rather a landscape I had inhabited my whole life. When I parked my car and strode across the sidewalk to ring Jaimie's bell, I stepped over the boundaries of that familiar butch terrain, gave away the compass, and lost my not-very-butch self. □

Judith Barrington lives in Portland. Her third collection of poems, Horses and the Human Soul, *was published last year by Story Line Press, Ashland, OR. E-mail: judithb@pacifier.com*

MOON OF MONAKOORA

Kevin Bentley

The middle age of buggers is not to be contemplated without horror.
Virginia Woolf

Well, at my age you can't lean against a palm tree and sing "Moon of Monakoora."
Dorothy Lamour, on her appearance as a sloppily dressed housewife who gets murdered in *Creepshow 2,* quoted in her obituary

*I*n the early eighties, I used to see this older guy dancing by himself at The End Up. He was *old*—white hair; wrinkled, leathery skin—and wore only a leather vest and a ripped pair of cutoffs. He'd be waving two large Japanese fans, entirely at odds with the disco beat. He'd stop now and then to open and inhale from a bottle of poppers, or to blow inanely on a whistle. Dancing Bear, we called him, snickering, recalling the stiff-legged character in a furry suit who'd come out periodically and shuffle wordlessly to some cracked polka on "Captain Kangaroo."

"Shoot me if I ever come to that, will you?" my friend Michael would say—pretty, boyish, shaggy-blond-haired Michael, dubbed Little Michael, because of his diminutive height and appearance on Polk Street when he was only 16. Or he'd croon in his best Peggy Lee manner, "She stayed too long at the fair."

"I hope to God I have something better to do by then," I'd say to myself. Something better meaning, presumably, a devoted lover, a real job, and a nice apartment with room for a big rolltop desk at which I'd sit typing out poems and stories, too happy and satisfied for drunken ennui in dance bars.

"I just hate it that you've chosen such an unhappy lifestyle," my mother would groan each Saturday, when she phoned from Texas. Marrying my childish father and bearing three sons hadn't been such a jolly lifestyle-choice for her, as it happened—my parents bickered, and each son disappointed, or worse. My older brother went to prison. I took drugs and fled to San Francisco to be a professional homosexual. My younger brother skipped the drugs,

179

but refused to go to college or stay married. My mother counted the days till she could retire from elementary-school teaching, where she hated the kids, and then hunkered down in misery with my half-mad father behind the bars he'd installed on the windows and doors of their ranch house.

My thoughts fall into elementary-school addition and subtraction: Twenty-seven years since I came to San Francisco; sixteen years since Jack died; twelve years since Richard died. Twenty-one years since I seroconverted.

The "older" men I went with in my first years out are now, if they're still alive, in their early sixties. I tote up their ages: 61, 65. I'm now older than either of my two lovers were when they died. They've morphed into younger men smiling out at me from the photos, as if they'd taken off on one of those light-years-long space missions and come back still in their primes to find me grown grandfatherly in gravity-driven earth-years.

A man I had an affairette with at 22, a handsome, urbane New York Jewish photographer, whose age awed me—39—used to say wistfully of the gay bike-messenger boys pogoing at The Stud on Punk Night, "They just don't *see* me. I'm invisible."

"What am I, a potato?" I'd ask myself, bitterly. Frank came to meet me once at the bookstore where I worked, dressed for a job interview in black jeans, a vest and tie, a vintage sports jacket, and a rather alpine hat. "Your boyfriend's out front," a woman I worked with told me. "He's cute. Very dapper." The word lodged: One could be middle-aged, and yet *dapper*.

On my first trip to a gay ghetto, in Dallas, Texas, at 19, I went home not with the hunky young truck driver beside whom I'd stood most of the evening (who'd muttered to me smuttily about whether I wanted to come out to his rig and try some of the things he'd heard queers were willing to do), but with a fussy, skinny, hand-cream-scented older gentleman who had swooped down on me as I started out of the bar. His house was a prettily restored Edwardian; glass tinkled from curio cabinets and tasseled lamps swayed like seaweed in a current as we stepped through the door and onto the echoing hardwood floor. I inhaled mildew and furniture polish. Were these the odors of forbidden sex?

He wanted the lights out before he whisked off his clothes. He dove under the sheets. I slowly pulled off my dingo boots and tight jeans and turned toward the bed. Standing in the moonlight that poured through the high, lace-curtained windows, dick sticking straight up, I trembled with excitement at my first *trick*.

"Stay there a minute," he said. "Let me just look at you. You're like a Beardsley satyr." He began to explain who Beardsley was, assuming I didn't know, and I accepted my role: young and dumb. His penis was small and only partially erect. Motionless, he ejaculated with a sigh, when I bent to take it in my mouth. "Sorry," he said. "I have a very short fuse." Then I lay back with my arms behind my head, while he blew me till I came.

Fifty: I picture gray-mustached men with that matronly roundness continuing below their belts, cheeks and noses ruddy with broken blood vessels, peering over their cocktail glasses behind the plate glass of The Twin Peaks (a.k.a. The Glass Coffin). Sure, Ned Rorem goes on looking great, but look what he started with.

You can get fat, or stop thinking of your choice of clothing each morning as "outfits," but if you never were a heart-stopping beauty in the first place, if you were never an Alan Helms or AMG model, you don't really have to worry about *losing your looks.*

For an oral history project, I once spoke with a 77-year-old gay man about his experiences in Nebraska in the late thirties. He told me about a cabal of "straight" pillars of the community—married doctors and lawyers—who used to rent hotel rooms in Omaha on weekends to entertain college boys—including my subject, Franklin—and street trade. The older men played poker and drank; the boys drank the free liquor and stripped down to their underwear in the summer heat. Late in the evening, the youths would have sex with each other on a couch pulled into the center of the room, while the men drank and watched, shouting encouragement, as if ringside at a fight. Then the hosts would each choose a student or farm boy or street tough with a pretty uncut dick and suck him off.

Franklin soon ceased to be *trade* ("This year's jam is next year's jelly—"), but he continued to introduce curious and comely young men from his circle to the weekend bacchanals. Certain "traditions" were insisted on. He was taught how to mix a proper scotch and soda. He was never allowed to pay for a drink or dinner. "Oh, no," the older gentlemen would say, covering his hand as he reached for his wallet. "You'll do that later."

1977: Attending a fundraiser dance at California Hall, I wandered with my pal David into a dark room where porn movies were being screened. We squeezed down an aisle and sat on a rickety bench. I hadn't yet seen much homoerotic film, and I was mesmerized. The person on the other side of me moved right up close. I scrunched nearer to David, hissing "Move down!" in his ear. The unwelcome suitor followed. I glared at him: an elderly Latino fellow, wearing his black windbreaker over his head like a mantilla. When his hand darted into my crotch, I leapt to my feet in outrage. "Keep your fucking hands off me!" I shouted, and pushed my way angrily out of the room. David caught up with me at the bar where I was shakily ordering a Bud.

"Do you have *any* idea who that *was?*" he asked. David had just started his first bartending job at a sad little Tenderloin dive called Googie's, and he'd undergone a crash course in Royal Court genealogy. "That just so happens to have been The Dowager Empress Manuel!" He was aghast at my breach of court etiquette.

"I don't care if it was Princess Matchabelli, he had no right to grab my dick!"

David, ten years older, gave me his patient, you're-so-young look. "You'll be old one day yourself, you know."

"Yeah, well, I won't be going around assaulting 21-year-olds and expecting them to put up with it!" I said irately.

When I was 18, I wrote a fan letter to John Rechy after reading *City of Night.* Rechy was from my home town, El Paso, and he graduated in the same class at El Paso High as my mother. I was six when, on our weekly visit to the neighborhood library branch, she hurried me past a display stack of *City of Night,* murmuring, "I knew him in school. He was strange." The jacket art showed a slender man silhouetted against a wet, midnight street, neon and car lights reflected in the puddles like Christmas tree lights in a poorly exposed color slide. My mother didn't say the word *homosexual* then, and I wouldn't have known what it meant, but the book and the image on its cover reached out to me. Later, when I bought and read a worn Grove paperback with an enigmatic phone number penned inside the back cover, I knew, despite the dated and grim scenario, I was reading about a place I meant to go. The phone number was disconnected when I screwed up my nerve to call. So I wrote the author, thinking he might be pleased to hear that one of his ridiculers, my mother, had borne him an heir; I enclosed some giddy poems. He answered with a friendly letter, informing me that he was hard at work on the screenplay for *City,* and inviting me to drop by his L.A. home should I find myself in the neighborhood. I saw him not long ago in an *Advocate* spread, still in his Black Bart garb—tight black jeans, open black shirt, and cowboy boots—muscular and proud, but fragile-looking, too, like the Diane Arbus shot of Mae West in her negligee. When will my Levi's and Gap T-shirts and high-tops render me as anachronistic as an elderly flapper?

Tradition provides a number of appropriate names for middle-aged gay men to use when addressing their juniors: Lochinvar, Youngman, Handsome, Apollo, Sport, Paolo.

"You're 37? I'm 32. That makes me the younger one—so I guess I'll be getting my way." This was uttered with a weary sigh and the first of many shrugs. We were lying on the rug in my living room, books, photos, and glasses of wine spread around us. In the breaks between our kissing and rolling around the floor, I was trying to talk Nick into having sex on this, our first solo date. "Of course, I'm attracted to you, but maybe this isn't a good idea." His show of reluctance, his insistence on the predictable sameness of all romances, ought to have alerted me, along with his conviction that five years' age difference gave him a perpetual trump, but the possibility of having sex again after four months of celibacy was irresistible. I knew from past experience that nothing helped put loss into perspective faster than a good dose of sex. I'd endured one very miserable year between Jack's dying and my meeting Richard, and I felt no compunction about skipping a semester at misery school this time. I didn't want to be like my friend Rick's neighbor Sal, whose new lover everyone referred to as Sea Monkey Boyfriend (recalling those add-water seahorses advertised in comic books), because Sal met and moved in with him only weeks after Andy died. Still, a week, a month, a year—when you've watched lovers die, when you're infected and asymptomatic with no idea when HIV may dissolve the earth beneath your feet and suck you under—what's the difference, really?

Three months and innumerable *Dark Shadows* videos later, when Nick, cigarette drooping from the corner of his downturned mouth, shrugged and lectured me one more time about sex being merely another bodily function, like going to the bathroom, quoting a maddening passage from French feminist deconstructionist Luce Irigaray that cast my penis obsession in a sickly light, I realized he'd been right: this wasn't a good idea. I walked away and steeled myself for the long haul.

It was a tough wait for the next good thing. Attending a revival of *West Side Story* on my 40th birthday, I wept when Tony and Maria sang "Tonight." It dawned on me then that for as long as I could remember, when I read about or saw young lovers, one of them was always me. But now an unpleasant little voice inside me was saying, *That's never going to be you again, honey.*

"I can't picture the kinds of things you say now, coming out of *his* mouth," my current boyfriend says of me in my twenties, after perusing my old photo albums. I'd packed all the albums and diaries into one big carton and lugged it over to his loft in the first stage of moving in. Having boxed up similar items for two dead lovers, the sight of my recorded transit gathered into one pile gave me pause. Because I've stayed put these 27 years in San Francisco, because I regularly leaf back through previous journals, for me the past and present are fluid, permeable. Other than the up-close arrival of HIV in the mid-eighties, there's no obvious demarcation separating young from young-no-more. I *am* that boy, give or take ten bongs and a six-pack.

Something better to do did come along, and I tired of drinking in bars or seeing how high I could make myself and still maintain, without having to bottom out and reduce my surname to an initial. The pre-AIDS diaries do seem engineered to excite and insult me now, in my more sober domesticity, with their sheer volume of sexual contacts and offhand slurs against those past 35 or 40. Yet I'm precluded from fretting about middle age by the very horror itself: How can I carp about birthdays, when every day I live, whole and healthy, cheating HIV out of its next meal?

Twenty-one years of asymptomatic infection, and the good news is: You may grow old.

Am I dapper yet? Do I drool over younger men? Will I Do That Later? Will I one day pass a poster for Spiritual Retreats for Gay Men Over 40 without laughing? Does childlessness, sexual voracity, and a fondness for small dogs invariably lead to Wallis Simpson? Can one be elderly *and* masculine without a past in the merchant marine? Will my tumultuous puberty ever end? If it does, will I have Something Better to Do? I intend to find out.

My forties are proving very sweet. I placed an ad and he answered it, responding, it seemed, as much to the series of more confessional texts I wrote and tore up as to the one that actually ran. Spit-spot; I moved in with him, after a year, leaving the apartment I'd lived in for six years, where my last lover died five years before. And 41 was a winning lotto number, the man behind door number three.

Little Michael, who died young and stayed pretty—I wasn't asked to shoot him, as it turned out; he did it himself when he started to get sick—came to me in a dream a few years ago. Putting his arm chummily over my shoulder, he told me about the rest of my life, smiling and teasing. "So don't worry," he said. "Of course, you know you can't remember any of this future stuff when you wake up. I'm not going to be coming back to talk to you anymore after this, either. I'm moving on."

He said it like he was heading to a bar for which I was neither attractive nor young enough. But he was laughing, affectionately. □

Kevin Bentley lives in San Francisco. He is the author of Wild Animals I Have Known: Polk Street Diaries and After *and* Let's Shut Out the World, *a collection of autobiographical narratives, both from Green Candy Press, San Francisco, for whom he also edited* Boyfriends from Hell. *E-mail: bentlek@aol.com*

ELK

Jon Remmerde

One summer, twenty-five years ago, I worked government contracts for blister-rust control in the forests of the Sierras in northern California. I uprooted currant and gooseberry bushes to remove an essential host for blister-rust, which kills sugar pines. I had hired on four helpers, mainly to keep my wife happy. She would bring our kids up and spend part of the summer in the mountains with me—if we had enough people, including a couple of guitar players, so we could party around the campfire after the day's work. Two of my helpers also brought their families up to camp.

Because our checks from the government took a long time to come through, we ran short of cash for groceries. I killed eight deer that summer and fed everyone with venison.

No one working with me would kill a deer. They ate the meat, but they wouldn't kill a deer. They killed ground squirrels for fun and left them where they died, but they wouldn't kill a deer to feed themselves. Their refusal to share responsibility angered me and added to the emotional turmoil that filled our camp. My marriage broke up, though I didn't know until the end of summer that Sharon was having an affair with one of my helpers, Don, and planned to unload me and marry him.

Before I knew anything about Don and Sharon, late in June, the first check came for work we'd completed. I drove down into the Sacramento Valley and cashed the check. I drove back up the mountain and pulled into camp after dark. In the light from the fire, I counted out what I owed to each of the crew. Then I said, "All of you pack your gear and get out of here tomorrow."

No one said anything but Don. Don said, "This is national

forest. It belongs to the public. I can stay anywhere I want."

I broke small twigs and fed them to the fire. Fire ate the small, dry fuel and burned bright and hot. I put the frying pan on the grill over the new, hot flame. I said, "Be out of here before noon tomorrow, Don. Stay anywhere you want, but not here. If you stay where I have to see you every day, you're inviting trouble. Anybody knows better than to invite trouble." I cut venison from the last hindquarter and put the meat in the frying pan. The meat sizzled against hot cast iron. The rich smell of searing meat stirred my hunger.

I cooked the last of that venison, and, from then on, I lived on groceries from the store. I worked alone, and I haven't killed an animal since that summer, 25 years ago.

Twenty-two years ago, I rode a BMW R-50 down from the Sierras on Highway 299. A drunk in a car slammed into me in the town of Montgomery Creek and busted me up. The driver had nothing, no insurance, no money, no property, no job. I lived through ten really rough years financially. I recovered physically as much as I ever would, but I still had enough problems that I couldn't work full time.

I became part of the mountains. Mountains became part of me. When I lived in towns in the mountains, I worked odd jobs for people who lived in the towns. I took care of lawns. I built gardens. I replaced broken windows, repaired plumbing, fixed roofs, built storage sheds. Away from towns, I worked government contracts. When I worked contracts, I was my own boss. I worked whatever hours I wanted to or needed to.

I met Laura and married her. The need for money and material possessions was no more part of Laura's consciousness than it was of mine, or we wouldn't have made it together. We raised two daughters, Juniper and Amanda, and started their schooling at home. They learned how to read and how to learn, and began to manage much of their own education with our support, love, and guidance. Eventually, Amanda and Juniper chose a year of public high school each, to see what it was like, and then headed off to college and into the world.

While they were growing up, I took care of ranches, and my employers paid me for the work I did, not for how many hours I could stay on my feet.

"We are wealthy," my younger daughter, Amanda, said. She

was 15 then, and she knew millions of people in this world had been driven out of their homes, off their land, had been killed, were starving. She said, "We are wealthy."

She was right. We had more than adequate shelter. We never went hungry. But, in the tightness of living on the edge, with a part-time job as caretaker of a Girl Scout camp, housing and a small wage provided, I became nervous about the future. Siren songs from the consumer culture built an insidious path into my consciousness. I feared lack. At least I could put meat in the freezer.

My older daughter, Juniper, 17, said she wanted to hunt. In the city at the foot of the mountain, she took the required two days of hunter safety classes. That autumn, we borrowed rifles, bought licenses, and walked through three inches of snow up into the national forest.

Fresh elk tracks in snow. More than 30 elk had traveled around the base of a bare granite formation standing high against the sky, then out across a meadow, through 300 yards of forest, and down into a rocky canyon.

Juniper and I stood on the rim of the canyon and looked across rough granite mountains. Trees grew in every available bit of dirt, among the rocks, even from the rock itself. Small meadows of grass, yellowed by autumn, white with snow, grew in flat areas. Aspen trees along the edges of meadows shed their last yellowed leaves and stood bare for winter.

I said, "Elk travel faster than people on foot, even when they're not in any hurry." I looked down the canyon's steep, rocky slope. "If we killed an elk down there, we'd do some heavy work to get it out."

The third day, the elk traveled back up from the rocky canyon, through the forest, across the meadow, and up the mountain. Juniper and I looked at tracks of the herd that had trotted west. Midday sunshine melted snow on the open meadow.

I said, "Tomorrow or the next day, if no other hunters stir them up and get them off schedule, they'll come back down."

The fourth day, Juniper and I sat all day on two points of rock, 200 yards apart. Snow fell from dark clouds, blew across the meadow, and dropped out of the wind when it hit trees in the forest above the canyon.

Late afternoon, snow fell thicker. I couldn't see Juniper anymore. The end of the day grayed the falling white snow. I climbed down from my point. When she saw me, Juniper climbed down from her rock. I asked her, "Cold?"

"My fingers and toes." We unloaded our rifles, pocketed the cartridges, and walked east, onto the ranch, rifles cradled across elbows, pointing away from each other. We walked down the road in darkness.

I said, "If we didn't know where we were, we'd be lost."

Juniper didn't say anything. She probably nodded. I couldn't see her well enough to know. Falling snow cut off most of the light that usually shines at night. Once in a while, we bumped shoulders, each not knowing, in the dark, where the other was.

Juniper spoke from the darkness, "We're pretty close to the junction."

I took my flashlight from my pocket. "Accurate sense of where we are, Juniper. This is the junction all right."

We started up the right road. I shut off the light and dropped it back into my pocket.

Laura had been worried. Juniper and I took off our outer garments and stamped snow from our boots. Laura started to chew us out for staying so long after dark in a storm. I said, "We got here as quick as we could. Snow slowed us down."

Juniper said, "Getting here late is better than never getting here, isn't it?"

Laura took a deep breath, as if she meant to say more, then let it out, hugged me, hugged Juniper, and walked back toward the kitchen. "Dinner's ready. We've been waiting for you."

In the morning, I said, "I have to do some work at the lodge this morning. I won't be able to hunt until afternoon."

Juniper said, "I'm tired of sitting on rocks and waiting for elk to show up. I'm gonna hike up the east side of the ranch and along the edge of the canyon. I'm gonna hunt above the north end, west of the drainage."

I didn't hunt until mid-afternoon. I thought about not going, but there were only two days left in the season, and I wanted meat for the freezer. Snow had melted and left open patches of ground. I walked rapidly until I left the ranch and walked into the national forest. Then I walked just off the edge of the meadow, through the trees, a slow step at a time.

The day was almost gone when I saw elk among the trees. They saw me. I saw three sets of antlers as the elk milled around, trying to get a better look at me. I looked through the scope at a bull, but it walked among the other elk until I wasn't sure if I held the scope on the bull or on a cow that had moved in front of him.

I stepped several slow steps and braced the .30-06 against a pine tree. The elk walked behind a granite formation, and I thought that was it, I'd missed my chance. But instead of running away, the elk went behind the rocks and came out on the other side. A bull stood in the clear, between two trees about 200 yards away, his left side toward me, his head turned, to look at me.

I put the crosshairs on the bull's head. The scope went black. Light faded too fast. These weren't the old days, when I always hit the animal in the head. I moved the rifle, backed away from the scope a little, and lined the crosshairs up behind the elk's shoulder. I squeezed the trigger.

Dusk on the quiet mountain, the .30-06 bellowed, punched my right shoulder back hard, and echoed from mountain to mountain. The bull rocked backward when the 180-grain bullet hit him. The rest of the elk turned and ran into the trees. The bull turned and followed. I had expected him to rock backward, as he did, then settle slowly to the ground.

Instead, he turned and ran. I slid the bolt back and ejected the empty casing, closed it on a live cartridge, set the safety, and let my breath out. I ran to where the bull had stood. No blood. I followed tracks.

Three hundred yards along the trail, the bull had fallen, skidded in the snow, then gotten up and run again. There was blood where he had fallen, but not much.

The bull turned, separated from the herd, took two smaller elk with him, headed north. I couldn't see the tracks anymore. I pulled the small black flashlight from my pocket. Tracks showed the bull was stepping lightly on his left front leg, heavy on his right. He and the two smaller elk had turned east.

I followed their tracks across the meadow, through the timber, to the edge of the canyon. The light from my flashlight dimmed to faint yellow. I couldn't see the tracks clearly anymore. My left knee, the one smashed in the wreck, hurt.

I walked back onto the ranch and walked the road home. Juniper had completed her day's hunting without success. I said, "I hit a bull, but I didn't kill it. I tracked it until the batteries in my flashlight died. I'll pick up the tracks at daylight and see what

I can do."

Juniper said, "I heard the shot. It was too close to dark by then." I didn't say anything. If the bull had dropped where I shot it, I could have bled and gutted it and skinned it enough to hold until I could hike home and get the pickup. The way it had turned out, Juniper was right, I had shot the elk too close to dark.

In bed, I tried to compose myself and sleep. I might have a hard day of tracking ahead of me, and I needed to sleep, but I lay awake a long time.

I woke long before daylight, fixed breakfast, and ate. I didn't wake Juniper. I drove the pickup as close as I could get to where I'd left off tracking the night before. I walked to the edge of the canyon. Morning's first light showed the rough, rocky slope falling away below me, with trees, grass, and brush growing in every patch of soil. The eastern slope had received sunshine and sharp wind. Snow had blown off and melted away and exposed open ground.

I followed tracks down the steep slope. They began down in the drainage. The biggest bull still used his left front leg very lightly, and the ground still showed no blood. I walked slowly and studied the ground. I turned to the steep slope so my weight came down on my stronger leg.

The ground leveled out. The drainage, with no water in it this late in the year, spread out into a small meadow. The three elk had headed north, across open areas of ground that had been frozen when they crossed. They left no tracks.

I walked up the steep slope to jumbled granite boulders the elk couldn't have climbed, then down the slope, across the north edge of the small meadow, to where flatter ground gave way to steep slope again, then upslope again, 50 yards north of my last traverse. Casting back and forth, I picked up the tracks again in snow, but then the snow gave way to clear ground.

I found tracks in snow, but again, there were too many. I couldn't separate out the wounded elk.

I walked at random from snowy area to snowy area. I found elk tracks—a single elk, a few elk, then 15 or 20 traveling together—but I never again found an elk favoring one leg.

The sun moved west over the edge of the canyon and left the eastern slope in shadow. There were hours of daylight left, but it would take me hours to climb out of the canyon.

The last day of the season, I walked along the rim of the canyon. Elk had walked up out of the canyon, but I didn't find tracks of a bull favoring one leg. A wounded elk, once it was down in the canyon, probably wouldn't climb the steep slope back up. I looked at the tracks anyway and walked along the edge more than a mile each way.

My knee hurt. I needed to stay off it for several days, once hunting season ended. Juniper walked up into the national forest west of the ranch and didn't come back until after dark. I asked her what she'd seen, and she said, "Fresh tracks, but nothing in them."

I had trouble going to sleep that night and every night after that for a while. I'd think I was going to make it, drift away into sleep, and I'd see the bull elk looking at me, then rocking back onto his rear legs as the 180-grain slug hit him hard, then turning and running, tan-colored rump disappearing into the trees.

I snapped awake. It was done with, and it didn't do any good to agitate myself, but my memories didn't ask permission to appear, they just appeared.

I dreamed I was in the forest in the winter, badly wounded. I was just trying to keep moving. I wanted to lie down in the snow and quit trying, but I knew if I did, I would die. Maybe that would be O.K., to lie down and die, but I couldn't quit.

I woke in the dark room, and my knee hurt worse than it had for many years. I propped my left leg on a pillow, because sometimes it locks straight, and it's a sweaty, painful job to get it to bend.

If I had it to do over, I would shoot the ought-six at least 20 times before I started hunting. I'd been tight with ammunition, because we had $70 invested in hunting licenses. We didn't have much ammunition, and I didn't want to have to buy more. I had shot the rifle three times and put all three shots in a four-inch pattern in the center circle, so I didn't fire it anymore.

I might have flinched, there at the edge of the meadow, when I sighted on the elk's shoulder and pulled the trigger. I wasn't used to the rifle, and I might have flinched, just as the firing pin snapped down against the cartridge, and I might have pulled the rifle off target.

When the scope blacked out, I should have let the elk go. Or I should have stayed with my original decision to shoot him in the head. A head shot would have missed completely or killed it. I had been too sure of myself. I had been greedy. I was stupid to revive my long discarded notion that wild species are ours to use for our needs.

There was so little blood, maybe I hadn't wounded it badly. Maybe it had recovered. But I didn't really believe that: A wounded animal facing winter doesn't have much chance.

I wondered where the elk was. I could know that kind of thing if I tried hard enough—if I just opened my mind and *observed* the world around me. When I was younger, I thought I might be able to do this sort of thing, but I was 51 now, and it had never happened, so it probably never would.

I wanted an end to thinking about the wounded elk, but I didn't know how to shut down the thoughts. I considered waking Laura, but let that pass. She didn't like to be waked.

I got up, dressed, walked out of the bedroom, and sat in the living room. Moonlight shone in through the windows. It was ten degrees outside. Wind blew through the forest, across the mountains, and down into the canyons.

Two storms, a week apart, brought snow. Juniper wanted to take the .22 rifle down by the creek and sight it in. I found ammunition for her. I said, "You want me to go with you?"

"You can if you want to."

She shot empty cans, and she hit right where she wanted to hit. She said, "I'm going to go see if I can kill a rabbit or something for the table." She walked up the road that ran beside the creek to the west boundary of the ranch.

I walked home in the afternoon sunshine. For a while, I sat at my desk, looking out the window at our part of the Rocky Mountains. Laura and Amanda had gone to town. They wouldn't be back until late. I decided to call my mother. I picked up the phone and dialed a thousand miles. We talked a few minutes. I started to hear what the Oregon members of the family were up to.

Juniper threw the door open. She breathed hard and said something I didn't quite catch. I spoke into the phone, "Just a minute." I asked Juniper, "What did you say?"

She took a deep breath, calmed herself, and said, "I killed a bull elk. I killed a four-point bull elk. Come and help me with it. You've got to come and help me with it."

I spoke into the phone again, "Mom, I'll call you back later. It might be quite a bit later." I hung up. I asked Juniper, "Where?"

"Not very far. Maybe a hundred yards. I don't know how far it is. Come on. I'll show you."

I grabbed my hat and scarf and gloves. "Are you sure it's dead?"

"It's dead."

"Wait. I'll get a knife. We'd better get right to work on it. It's going to be dark before long." I got my pocket knife from the desk drawer and checked the blades. I knew they were sharp, but I checked them anyway. Juniper leaned the rifle in the corner and we went out and shut the door. Juniper led me down through the trees, down the slope north and west of the house.

About a hundred yards from the house, the elk lay dead, head downhill beneath a small tree. I took my knife from my pocket, opened it, and dropped to my right knee next to the dead elk. I slit the skin up the throat, peeled it back, then sliced between muscles until I found the carotid artery and severed it. Some blood came from the artery, but not a lot. The bullets that had killed the elk had largely bled it. The snow under the dead bull was red with blood. Juniper said, "He was wounded. His left front leg is gone."

"I saw that. What happened?"

"I heard noise up in the rocks, so I stopped and listened. He came down out of the rocks and stopped and stood sideways to me and looked right at me. He knew I had the rifle, and he wanted me to shoot him. I shot him in the neck, but I was too far away for a .22, and it didn't kill him. He started to run. I caught up with him and ran beside him and shot him again, right here, high in the neck, and he fell, but he wasn't dead. So I shot him in the head. I waited a little while, to be sure he was dead. Then I came and got you."

I stood up. "Let's get him down onto the meadow. Let's get the pickup. We can drive across the meadow to the base of the hill. We'd better move. It's going to be dark soon."

Juniper brought the pickup to the shop, while I gathered rope, a saw, and a come-along. We drove around by road and across the meadow and dragged the elk out into the open.

Juniper had worked with her cousin, Jason, on a deer carcass, so she knew some of what to do, and I gave her instructions. "Skin this side back. We want to get the hide back far enough so we can cut the quarters free. Watch while I cut so we can get the intestines out. The knife stays shallow, like this, so we don't cut any of the guts and make a mess. Keep the blade between two fingers, so you press the gut down out of the way as you cut. Now pull all that out onto the ground."

She reached in and tried, but she said, "I can't get it out. It's too heavy."

"Let me have a try. That is heavy." I leaned backward and pulled the intestines out onto the ground. "O.K. We need to separate out the liver and the heart."

We cut the hindquarters free, cut the body into two pieces, and loaded all the pieces onto the hide in the pickup. I said, "We can come back in the morning and get everything we're not going to use and take it up to the ranch and leave it for the coyotes and ravens."

"I want the antlers."

"Here. Use the saw and cut this way and this way until the two cuts join and then pull the antlers free."

The day's light faded. Moonlight and starlight reflected from clean snow. When she had the antlers free from the head, Juniper stood up and said, "Maybe this is the one you wounded."

"Could be. He could be the one."

"That's what I thought when I first saw him, that it was the one you hit, and he wanted to end his struggle. He wanted me to shoot him and get it over with."

"He had a rough time of it. The hide and hair grew over where his leg was gone, but he still had a rough time. His paunch is full, but he doesn't have any fat left on him. Hard for an elk to make it with one front leg gone."

We hung all the parts in the shop. I said, "I'm not going to be able to let the meat hang as long as it should. Too much chance somebody will come along and see it. This long after the season, we'd have some trouble."

"I would. I'm the one who would be in jail."

"I suppose you would. We could call Fish & Game and explain what happened. Killing an injured animal that was having a hard time making it, they probably wouldn't bring any charges against us."

"Would they let us keep the meat?"

"I don't think so. I think they'd take the meat."

"Then we're not going to do that. I killed this elk, and we're going to eat the meat. I didn't kill it to give it away."

"O.K. Let's go get cleaned up. I'll cut it and wrap it and put it in the freezer tomorrow, while you're in school."

When Amanda and Laura came home, Juniper told them. She said, "Don't tell anyone at all, unless you want me to be in jail."

I stayed up late that night, after everyone else went to bed. I thought about what Juniper said, that it must have been the bull I hit. Maybe it was, but I didn't think it would have come back up out of the canyon if it was wounded.

Then I thought, Jesus, I'm tired of being a logical, reasonable adult, with all the attendant details and problems and hangups and restless thoughts and restless dreams. I really have no idea what any elk would do.

I shut the lights off and walked downstairs. I got ready for bed in light from the moon. Laura snored. I got into bed without waking her. I lay there looking at trees and sky and the moon.

When someone gives you a gift, accept it. Don't turn it around and around and dust it off and polish it and inspect it and try to decide if it's perfect. Because it is a gift, given freely and with her total being. A perfect gift.

I put my hand on Laura's hip. She moved toward me and turned onto her back. Her snoring changed to deep, even breathing.

Troubled thoughts were on their way out. There wasn't time in this life for dark moods to stay with me. I swam up out of darkness into brighter country, where each day was only that day, with nothing before it and nothing after, and I began to live in that brighter country. □

Jon Remmerde lives in Bend, OR. He has published five books with iUniverse, including a novel, two collection of stories, a collection of essays, and Somewhere in an Oregon Valley: One Family's Adventures Taking Care of a Ranch in Northeastern Oregon's Blue Mountains. *E-mail: remmerde@remmerde.com*

THE 20-BREATH SNAKE

Russ Riviere

*H*ook liked to tell this story: So there we were, on the headwaters of the Rio Custapec, above Finca Yaeger and a mile or so from the end of the road. Out of those deep canyons grow what must be the tallest and fattest liquidambars on earth. This was going to be the last collecting day at the last collecting location…the last day before I could get the fuck out of Mexico. After three months, I really missed my kids. I was also close to starvation. Alushe might be able to make it on the fare our boss, the Forest Pig, provided, but I myself needed burgers.

We, that is, Alushe, the Forest Pig, and myself, had just left the Mexican botanical team to its business of making breakfast. Quite the breakfast. They were frying eggs, they had fresh rolls, they had somebody's canned Canadian bacon, they had an assortment of fresh fruit, and they had some scornful snorts out of us. When you work for the Forest Pig, you don't need no stinking luxuries. We had already eaten. A couple of oranges was all we needed. Supplemented by the last of my personal stash of *animalitos*. They are the animal crackers you buy by the kilo. They taste like the Sunday funnies, but they have made the difference more than once in the minimum-requirement-of-blood-sugar department. Anyway, we had the jump on the Mexicans. That meant we could score the best plants while they indulged their weaknesses. I am certain that the Forest Pig had invited them on his collections only to humiliate them with his superiority. "Outcowboy the Greaseballs" was his philosophy in the field, and probably everywhere else as well.

Now, it was typical that Alushe and me would start each new day with our ears laid back just like any other abused beasts of burden. This morning was a little different, because, like I said, it was supposed to be our last. By this time, Alushe was almost as capable of climbing trees as myself...or at least of collecting in them independently. So we separated. The Forest Pig went up a ravine into the high limestone cliffs, on the make for ferns and orchids, Alushe headed off along the banks of the upper road, and I myself took the lower side of the road where I could see more clearly into the crowns of the trees. Steep slopes, you understand. I figured I had seven or eight climbs in front of me. There were more or less quarter-mile distances between us, but we were easily within shouting range.

At about 6,500 feet, it was kind of chilly, and I had been assured by the Snakelero, who had stayed behind in San Cristóbal, that we would be outside the range of all possible arboreal *venenosas*. At that time, I had great respect for the Snakelero, so I waded up the trees that day with a will. And will is what it took after the fourth climb. I was exhausted. Not only by the day, but by all the other days and by too little sleep and too little food. I was sick to death of carrying all my equipment around, and I was sick to death of no gratitude from the Forest Pig. So, from a slightly mitigated bad attitude I regressed quickly back to my normal, bitching, swearing self. What was worse was the nobody to share it with. Except another *pinche* tree. But there was no choice. Just get it over with.

There was this little scree slope that ran about 50 feet down to the object tree from the trail, and I figured I could negotiate it without uncoupling myself from my equipment. I probably could have except for the exposed root that caught my right hook and sent me slamming backwards into the trunk at what was probably closer to 30 than 20 miles per hour. That was bad enough. After I caught my breath and looked up, I realized how much worse it really was. The tree was a lot bigger here than it was from the trail. Plus it was completely smothered in epiphytes, aeroids, and bromeliads for the first 70 feet, at which point—the first crotch— it was still four feet in diameter.

I could have cried. Maybe I did. The pain in my elbow was even more intense than my hatred for science, but neither one was as bad as not wanting to climb that tree. However, when it comes down to it, everything else aside, I am a treeman. Unless there is a bar handy, I will climb the sonofabitch.

Now, climbing your typical rain-forest tree is not like the flip, chunk-chunk cleanliness of your typical temperate tree. This is not to say that the Douglas Fir, for example, is not without its frustrations in terms of hard jutting stubs and body-shredding bark, but in the scale of human misery, the rain forest tree is supreme.

There was no possibility that I was going to trust free-climbing the vines to the first comfortable tie-in, because I had experience in that technique. Normally, it was the basic ant attack that made you lose your hold in desperate situations, but there was also the likelihood that *guitarones* or other murderous wasps would be on hand for ambushes. Not counting spider ants or scorpions.

So I resigned myself to the long, but surely safer, technique of removing the vegetation as I climbed. This is done in sections. Overhead. Every three or four feet I would hack a ring around the trunk and rip the vines free. I don't know how to make this sound as difficult as it is. It is also hell on the habitat. Caustic sap is worth mentioning here. Some trees have it, some vines have it, and it can, in different and varying ways, bring great grief to men. Blindness and shrieking rashes are among the possibilities. And, of course, cutting anything overhead makes me nervous.

So there I was, hack-hack, voopa-voopa, chunk-chunk, rip-groan for maybe 45 minutes of sweat-straining agony.

I had gone about 60 feet when I came to the bell—the diameter of the trunk went from about four feet to about seven feet. I was now convinced that I had not checked this tree out in a thorough manner. In the meantime, I had buried my climbing line under 20 feet of chopped vegetation, and it was hopelessly tangled. I had to let it go. Having to let go is an interesting phenomenon. All of us have probably experienced it. Letting go of the climbing line is not as bad as letting go of the wheel, which is fairly sudden, but it is almost as bad. However, there is a kind of purity in being reduced to nothing but a waistline.

Looking up, I got a pretty good shot of the crotch, which to my further discouragement was hairy beyond the normal. In order to throw my waistline through it and make myself secure for the upper ascent, I would have to clean the crotch out. The only thing I could do was to climb as far above my waistline as possible and hack at the crotch with my machete. This was a maximum-strain situation that would surely turn into one of minimum effectiveness. So I did the easier thing. I poked around in the crotch the way border guards in French Revolution movies look for the hero in the hay wagon. Satisfied (if that's the word), I then unclipped myself from the waistline. This left me clinging to a vine with my left hand, while my hooks were tentatively set in the trunk. I then swung the waistline with its heavy clip (I have knotted my head more than once with it) through the crotch. All I had to do now was to inch my way up one, two feet, reach around the blind side of the trunk, grab the clip, breathe with relief, core in, and continue. The inching up was successful enough, but when I reached around the trunk to grab the clip—Whang!

Shit! Another suck-ass wasp, I thought, forcing myself not to overreact. But when I looked at my hand, I saw the trickle of blood. Blood? I looked at the crotch, then at my hand. Then at the crotch. Blood? Then I saw it and can see it yet. Slowly, lazily, preparing itself for the next strike—the 20-Breath Snake—the almost mythical grinning little pit viper that instantly paralyzes birds, frogs, and rodents in trees, and that had just bitten me!

Now I have spent a great deal of time thinking about this very snake. The Forest Pig had told me more than once, and with what he passed off as humor, that if bitten by one of these guys you'd have approximately two minutes to live. Which was why the TselTal called it the 20-Breath Snake. There was no anti-venom for it, and even if there was, I knew we didn't have any. We were also 200 miles from the nearest airport, which would be fogged in anyway. Added to that was the fact that I was 70 feet up a tree on the uphill side, 150 on the downhill side; not tied in; and I still had to deal with this snake.

Decisions had to be made. Should I kill it with my machete and take the chance its upper half would fall in my lap? No. Even if the head missed me, I wouldn't want to go to Hell with no snake. Should I flip it onto the trail so that Alushe and the Forest Pig could see what killed me? I knew that the snake had a prehensile tail and couldn't move well on the ground. No. I would have to get too close to it. I remembered that I had paid up my life insurance and was gratified in a strange way. I thought all these thoughts in the amount of time it took me to yell, "SNAKEBIT!" at the top of my lungs.

Stay calm, keep your heart rate down, get out of the tree before you fall out. That was my advice to myself. I backed down slowly, keeping my eyes on the golden, sloe-eyed ones of the snake.

After I caught myself from barking out any further, I took

another look at my hand. It was already swollen. I sucked the blood off. What precious little there was of it. I expected that dark blood you see in punctures, but instead there was this short pink scratch, not your direct, two-fang hit. I began to smell hope.

If anyone asks you what hope smells like, you can say it smells like burgers.

A couple of dozen pure, calm steps and I was at the base of the tree and also at the bottom of the scree slope I had fallen down before. I probably should have stayed there, but I couldn't. I mean, I had to reach the road. So I did, charging all-fours up the slope.

As I reached the top, black fingers began to run through my brain. I started to pull a strap off my pack for a tourniquet, when the Forest Pig came charging out of his ravine. I instantly forgave him for all his asshole ways. If I had to die, I couldn't find better or more profound company.

"WHERE'S THE SNAKE?" he boomed, as he skidded by me, out of breath, eyes furiously darting. This was not the question I expected, but caught up in the moment like I was, I pointed to the crotch of the tree.

"You mean you didn't get it?"

I groaned; he grunted: "Where did it get you?"

I raised my hand.

Then Alushe came rushing in; the Forest Pig sent him to get water from the creek, the Pig himself being busy searching for the snake through field glasses.

It was still there, writhing on top of the foliage remaining in the crotch.

"*Negroveridis!* By shit, it's *negroveridis!*"

The Forest Pig looked at me to share his enthusiasm.

He saw how far that was going.

"Don't worry. If you're not dead by now, you won't be. At least I don't think so. The problem now is how to get the snake. We have to have the snake!"

Alushe returned with the water, while the Forest Pig paced back and forth, building himself into that state of irreversible despicability that so marks his kind.

"We cannot let it defeat us!" he kept saying.

I soaked my hand, absorbed, as you can imagine, with the growing realization that I might be O.K., but I was not yet completely out of the action. I tightened my jaw and repeated no way, no way, to myself as the Forest Pig sidled up in his charmingest manner and said, "Take a close look at that sucker."

I accepted the glasses. I pretended to possess some kind of scientific interest. I did not say, "How the fuck much closer can I look, you prick?"

"You know, I suppose," the Forest Pig went on, paying no attention to my silence, "that there are only three specimens of *Bothrops negroveridis* in the world, and no live ones?" Of course I did; we'd talked about it plenty of times. What else do these pricks have to talk about? I still didn't say anything, because I was absorbed by the snake. I struggled one-handed with the focus, hoping to see the snake's magnified eyes overlay my memory of them closer up. It wasn't much use. My steadiness was shot.

"I'm glad I wasn't wearing gloves," I said, mostly to myself. I must have moved my hand the precise moment the snake struck. Had I had gloves, I would not have known that I had been struck. I would have inched my body into the crotch, at which time I would have been hit in the throat. Not a pretty picture. It occurred to me that something fatey was going on. Some creeping affinity.

"How are you feeling?"

I eased the glasses down and turned my head to look full into the eyes of the Forest Pig. How could so large a man have such glistening little rabbit turds for eyes?

"Grateful."

In reality, my left hand was beginning to look like a catcher's mitt, my legs were rubbery, and the rest of my body was frozen with uncertainty. The real deal was between me and time, and there was nothing the Forest Pig could say to encourage or encroach upon my attitude. I was in a state of grace between the living and the dead, and I fully planned on playing it to the hilt.

No slouch at reading between the lines, the Forest Pig took back the field glasses. He looked into the crotch for a long moment. He said, "You know, we're dealing with a disoriented animal up there. He's a night hunter who has been disturbed in daylight. It's cold and he has stiffened up. Probably why he missed you. Look at him, still on top of the nest ready to repel another invader. An extremely anxious situation for him, don't you think?"

"What makes you think it's a him?"

"There's only one way to find out, isn't there?"

We looked at each other dead on, and then, as if we had

rehearsed the move, both of our heads turned toward Alushe.

Now, Alushe has had long practice in looking ignorant. He can look irreproachable if he wants, with his classic Mayan features, or he can look like he doesn't know one end of a shovel from another. In fact, I've never known anyone with so many ways of looking ignorant. But this time, he knew that we knew and that there was no honorable way out. Especially since the Forest Pig held the money.

The Forest Pig began the dialogue: *"Baak I Laal."*

Sly. Real sly to call Alushe "Big Brother." To use the term that signals that more has been shared than ordinary relationships can bear, and so if you cherish me and know me and value our future friendship, you will respond.

"Baak I Laal, we have to have the snake. Hook cannot do it, and I carry too much weight. You can do it and you have our help. Hook now knows the name of the animal, and I will watch through the glasses and tell you how to move."

Effective. The Forest Pig may be a fucking jerk, but he is always an effective fucking jerk.

"Baak I Laal," Alushe began in the formal oratorical style, "Baak I Laal, you say that we must have the snake. I am not sure but that the snake already has itself. Hook has no cannot. He has done it. You say that you are too heavy and cannot do it. That is true. It is not what you are for. Hook knows the name of the animal and that is also true, but he cannot tell. You say that you will watch me, but you cannot be me. I am not ashamed to say that I do not like snakes and that they do not like me."

After what followed—a good half an hour of watching the Pig throw rocks at the snake—Alushe and I were not surprised when he finally said, "We'll camp here." Out of breath as he was, there is no end to his snot.

"Not only do we not have the snake, we don't have the tree."

Although we were looking forward to the return trip, this new decision wasn't as bad as it could have been. The afternoon was cool and sunny, every detail was crisp and golden, and my appreciation for everything around me was immense.

We made camp quickly and with few words. I wasn't much help, but the small tents went up easily, and as Alushe built the cooking fire, the Forest Pig changed tack. I supposed that my high spirits were infectious and that the Pig's change of attitude was simply an emergence into common decency. Sure.

Our meal consisted of one can of fava beans, some dried mountain tortillas called *kosh osh,* and what was left of the animalitos. We squatted around the fire and ate with slow bites and darting eyes. The way the Donner Party must have. The Forest Pig was in the habit of counting mouthfuls so as to make even the lightest eater feel guilty. He rationed tequila the same way, but that night, as I say, was different. The food game didn't change, but he positively plied me with tequila. And I positively drank it. He spoke of higher responsibility—how knowledge is only gained by sacrifice, et cetera. He told me what a neat guy I was. I guess that is what finally got me. With Alushe already snoring, I agreed to go after the snake the next day.

When I finally went to sleep that night, I dreamed clear, weird dreams. I will tell you here that there is nothing I can't stand as much as hearing somebody else's dreams. I would rather listen to their back problems. Mine are usually no different, but over coffee the next morning I told my dreams to the Forest Pig. No toothless old bag at Delphi ever got cloudier and uncloudier over a potentially fat customer than the Forest Pig got over me. No trench chaplain ever shoved a doughboy more eloquently out into no man's land.

I would have gone anyway.

"All right. Don't worry. I'll have my field glasses on the crotch every second. All you have to do is flip him out. I'll take care of the rest."

I returned this short set of Forest Pig instructions with a slit-eye nod. Alushe had taken a seat on a rock to watch. He turned his head and gave me the mock, sideways spit. We use this as a code when something is genuinely the shits. It helps to buck us up. I agreed with him, and it didn't take no gesture to see it. I headed back down the groove in the scree my body had made the day before. Yep. Same tree all right. Same amount of nothing on it.

I would only have to scale the mound of drying vegetation that I had stripped from the tree in order to arrive at the clean trunk. It would take me less than two minutes to arrive at the crotch. Great. I started up. As I ascended, I became sorrier and sorrier. At the same time, the Forest Pig, looking on from the safety and security of the ground, started revealing his true nature by making faces. The closer I got, the more intense and slathering and raw his faces got. All of my hatred for him reemerged. As I neared the crotch, my progress slowed to almost nothing. I might as well have had a Studebaker sedan tied to my climbing belt. Sweat was pouring in my eyes.

"YOU'RE O.K.!" the Pig yelled. "NOT A THING MOVING!"

Including me, I thought, giving him the searingest look I could imagine. I let out every inch of slack in my waistline. The snake could only strike out half the length of his body, you see. But even that was no use. I just couldn't force myself to get any closer.

"I'M GONNA TRY FOR THAT CROTCH," I shouted at the Pig, who always pretended to understand. What I meant was that I would throw my climbing line through a higher crotch, apparent now in one of the upper branches. I would then climb the rope and come down on the snake. Which I would have done the first day had I not had to deep-six my climbing line.

I bundled enough rope to make the throw and enough to fall back down to me once through the crotch. Maybe a 20-foot throw. The first shot missed. That is the rule.

I swore softly, re-coiled the rope, and got ready again. My concentration was intense. The rope arced up, missed the target, and came down through the snake crotch. Exactly the worst place for it. I would have to pull the rope free. Which is to say, it was now possible to pull the snake into my lap. My mood deteriorated.

"YOU CAN DO IT!" the Pig called, recognizing the situation.

"*YOU* CAN DO IT!" I called back, among other things.

Now, it doesn't pay to lose your temper at times like these, but I couldn't help it. In my rage, I thunked around the trunk, getting as far away from the retrieval as possible. I drove my hooks into the tree as hard as I could. I also drove my right hook into my left Achilles' tendon. Right up to the hilt. Through my boot and an inch and a half of cold steel further. I yanked it out just as hard. In no way did I want the Forest Pig to know. The sweat burned my eyes and my boot filled with blood. I lost all reason. I charged into the snake crotch...through it...then above it. I cored in with my waistline and flipped my climbing line into the crotch I had missed before. The Forest Pig was screaming, but I wasn't listening. I dropped back into the crotch. I ripped into it with my boots and my machete.

No movement. Not one sign of life. The snake was gone.

The anticlimax of it all made jello out of my legs, so I straddled the crotch like it was a fat pony and I watched the ants try to find a way into my pants. "He's gone," I said, more to the tree than to the Forest Pig. But the Forest Pig already knew. I half expected to see him throw the field glasses to the ground and stomp on the lenses. I collected a small branch to represent the tree. Fruits, flowers, leaves. Alushe came to the base of the tree to retrieve it. I climbed down.

We loaded our equipment into the truck. Nobody said a thing. My boot sloshed a little with blood, but it didn't leak. I was somehow pleased that my hand hurt more than my heel. I climbed into the back of the truck without even removing my hooks and collapsed. At least I was on the way home.

"Just one more stop," the Forest Pig said after we had gone about two miles. Nothing he could say could really get to me anymore, but I did wonder how far Alushe and I could get in the truck before the Pig's body was found. We pulled into a small turnout at the edge of the German coffee farm. Probably quite a ways.

Alushe and the Forest Pig gathered up the machetes and collecting sacks. I stayed in the back. My heel had stiffened up, and I knew that once I took my boot off, it would be a long time before I got it back on. I pulled my body out of the truck.

"It's not far," the Forest Pig said, some small endemic plant that he knew would be in bloom. Alushe and I followed as the Pig made his way up the trail. It was a good trail. Wide, clear, and obviously used as a kind of *periferico,* or perimeter trail, by the coffee workers. The first mile was bad enough, but after that, each step was torture. I got dizzy. I sat down. Alushe knew right away that something was more wrong than usual. Nobody sits down in the jungle. At least, not me. We agreed that I would wait for them to return.

It was a good place to rest, right beside the greatest strangler fig I have ever seen. It was a giant macramé cathedral. Completely hollow inside, the tree reminded me of some rich and beautiful widow who made her home on the decomposed ashes of her long dead husband. I fell asleep.

I wish I could say that when I woke up everything was different, but it wasn't. Alushe prodded me with his walking stick, so I had to begin the long walk back.

The Forest Pig took the point the way he always does when the trail is easy: Botanical Adventure Man, thrusting his way one perilous step further...into the unknown. Asshole. Alushe, behind me, was making shushing noises to remind me of my mumbling, when I saw the Forest Pig lift his elephantine leg over a branch that had fallen across the trail. What struck me was a small flash of electric green.

"Hey, uh, Jefe...this almost looks like..."

And sure enough it was. *Bothrops negroveridis*. The 20-Breath Snake. A teenager. Maybe it fell out of the tree. Maybe it crawled out onto the edge of this dead branch and the branch had broken. But here it was. Some kind of mystical gift to science.

The Forest Pig did a little dance around the branch. Dumbo around the china jar. Alushe was only slightly amused, but I myself was outright amazed. The snake was small, about nine inches or so, and the big problem was how to get it back to the truck. As innocent as he looked, he was not the kind of guy you drop in your pocket. The Forest Pig solved it. He broke off the part of the branch the snake was coiled around and walked it out of the woods. Or almost out of the woods. He had to stop every couple of hundred feet and turn the branch around, because the snake would wind his way toward the Pig's hand. This became a kind of comedy contest, and I knew who I was rooting for.

We were in sight of the truck when the snake dropped off the stick. It was like he had waited for the grass. The grass was just his color. He was not supposed to be able to move effectively along the ground, but then I had just received a big lesson in supposedlies. The fact was the snake was lost and I was glad. The Forest Pig ranted and rooted around for half an hour, and then Alushe, of all people, found it and flipped it onto the road. Easy pickins. The Pig got it into a jar and I rode all of the many hours back to San Cristóbal with this snake between my legs.

The only living 20-Breath Snake in captivity died two months later in the San Diego Zoo. Wouldn't eat. A month after that I was able to get my boot on. □

Russ Riviere, a tree man, storyteller, and activist, died last year. He taught himself Sanskrit and was a scholar of Western lore and fact. He did three tours of service in Vietnam as a member of the the 101st Airborne. A genus of oak, Quercus riviere, *was named in his honor.*

Printer's Ink In My Veins

Peter Booth Wiley

*T*here is a touch of printer's ink in my veins, inherited from five generations of my father's family. My great-great-great-grandfather, Charles Wiley, was really a printer, as no distinction was made between printing and publishing in 1807. Charles published James Fenimore Cooper's first truly popular novel, and his son, John Wiley, who gave the firm his name, published the first works of Hawthorne, Poe, and Melville. But from the little we know, we suppose that John was a failure as a literary publisher. Or perhaps writers steered clear of him after he censored Melville's *Typee*, forcing him to remove critical comments about the South Seas missionaries John Wiley very much supported.

Influenced by a son who was an engineer, John decided that "scientific" publishing made sense in a rapidly industrializing America. At the time of his death in 1891, only a vestige of literary interest remained, the continuing publication of John Ruskin. My father joined the firm in 1932 and became the fifth descendant of Charles to head the company. Two years ago, I became the sixth.

My brother and I put out our first publication when we were ten and twelve, after our parents gave us an old Royal typewriter. We painstakingly typed up one copy of a family newspaper, *The Wiley Herald*, struggling to give it a masthead and to keep the articles in columns. I have a vague image of a lot of straggling, typed-over words, some with broken letters, but can't remember what we covered: what we had for dinner? the activities of Barry, our dachshund?

In high school I accepted the job of sports editor on the

school newspaper, partly because of the camaraderie of the small office where we could work without interference from faculty members. We were also permitted to leave campus—after signing out at the front office, of course—for regular trips to the printer in downtown Elizabeth, NJ, and a stop at a soda fountain on the way for an Awful-Awful.

When we were younger, my father had taken us to a printing plant, and I was fascinated by the linotype machines, where long, gray lead bars were lowered into a melting pot, and molten lead flowed into molds and appeared as type, slugs of which we took home. The presses themselves were simply overwhelming, huge rolls of paper screaming through web after web to emerge as the printed page. In Elizabeth, I would stand mesmerized, watching the linotype operators bang away, sniffing up the odor of hot lead, not knowing, of course, that they were frying their brains. Printing was alchemy, the way the insubstantial stuff of one's mind found its way between two hard covers.

When I thought of a vocation, I considered publishing for a quick minute and then fastened for a time on the ministry for a curious and timely reason. I went to a high school religious conference where I heard Yale chaplain William Sloane Coffin talk about the Freedom Rides in the South. Coffin was a powerful presence, and I wish I remembered what he said, but I don't. All I know is that he unsettled my mind. I had been brought up an Episcopalian of a middling type, attending a shabby, barn-like church in the small town where we lived. I took religion seriously. I sang in the choir and then became an altar boy, struggling to make it through the long communion prayers on my knees. Coffin suggested another kind of religion, one that responded to the disturbing events that were taking place outside the self-absorbed world of suburbia.

I had seen the worst that America had to offer on my uncle's Long Island farm. I worked with him one summer picking and grading string beans and helping to manage a crew of black migrant workers. For a number of years he hired the same crew, because unlike other farmers he provided accommodations, a two-story farmhouse. Some lived in the house, others in their cars or in or under their trucks. This crew came back to work for my uncle year after year, because these accommodations were better than the alternative. The rest of the migrants lived in camps—one for blacks, one for Puerto Ricans—which were little more than sheds surrounded by the naked earth. They had that concentration-camp look about them. I had absorbed some strong moral lessons from Christianity only to find that the world didn't work that way.

From a suburban cocoon, I unwittingly went into deeper isolation at Williams College, nestled in the hills of western Massachusetts. But even Williams, then a venerable drinking establishment with bright students but few intellectual pretensions, was touched by the larger social issues. Inspired by an English instructor, I began to read *The Nation*, *The Partisan Review*, *The Realist*, and the left press, and then the Marxist and neo-Marxist classics and books on Cuba, China, the Soviet Union, the labor movement, utopian communalism, the economy. I probably spent more time reading on my couch than I did preparing for and attending classes. It certainly showed in my mediocre grades.

In the spring of 1961, the glory days of the Kennedy administration, a handful of us went to Washington to demonstrate against nuclear testing. We even marched down fraternity row, a pathetic little band made up of students and faculty wives and jeered at by the frat boys. We didn't care.

Under the leadership of the captain of the football team, some of us were circulating a petition to abolish fraternities, which is what ultimately happened. Another handful of students organized a support group for the civil rights movement. And as the war in Vietnam escalated, we picked up bits and pieces of information from the left press and, finally, when they began to cover it, from the established media. We also carefully scanned student newspapers from Harvard, Michigan, and Swarthmore to see what was happening on those, famously active campuses.

I struggled with my academic writing. Papers were torture. I would slave away, barely able to produce the requisite number of pages, all of them written in a ponderous style that I picked up translating Cicero in high school. The one time that I wrote freely was in a course given by a British Labor Party intellectual. We were asked to write a paper on Hannah Arendt's *The Origins of Totalitarianism*. I quickly dismissed her work and launched into a free-flowing exegesis on the relationship between traditional Marxism and the national liberation struggles in Cuba and Vietnam. My efforts were not appreciated.

I tried writing for the college newspaper, doing a long piece on faculty attitudes toward the House Un-American Activities Committee. I knew about the anti-HUAC demonstrations in San Francisco, and a group of us had invited Congressman James Roosevelt to the campus to talk about his campaign to abolish HUAC. Fewer students than could fill a small dormitory living room showed up, and I found the faculty to be strangely reticent and obtuse when I interviewed them. But even writing for the school newspaper was more than I could handle. I wrote a couple more pieces and then settled into a copy-editing job. My struggle with writing had become so bad that I even accepted a kindly roommate's offer to draft a paper for me.

In my senior year, 1964, the same roommate, a fellow radical, which is what we called ourselves by then, was the editor of the paper. We were faced with what to do when it was announced that Secretary of State Dean Rusk, a fervent supporter of the war, would be presented with an honorary degree at our graduation. There was no chance we could get enough people to demonstrate, so we staged a demonstration...in the paper. We made some picket signs and photographed three or four graduates in their robes marching in front of our dorm...and ran it on the front page! We all left town the same day, never looking back. □

Peter Booth Wiley was born in Orange, NJ, in 1942, and is now a journalist living in San Francisco, as well as chairman of the board of John Wiley & Sons. He is the author of The National Trust Guide to San Francisco *(John Wiley & Sons, 1999). E-mail: pwiley@best.com*

DEUS, AMICI, ET NOS

Eve Pell

*T*he family tree tacked up on a corkboard in my office goes back some 20 generations to a Walter de Pelham in 1294 in England. A much later Pell arrived in the New World in 1635 and bought, from five Indian sachems, a huge, vaguely defined tract of land in what is now the Bronx and Westchester County—the 1654 treaty was reenacted in 1956 as part of Pelham's 300th Anniversary.

The family's fortunes plunged after the American Revolution because most Pells were Loyalists; however, in the 19th century, one unusually enterprising Pell got rich on Wall Street—importing marble and fine woods. Several of his handsome, financially inept descendants maintained their prominence and wealth by marrying rich women.

My cousin, Claiborne Pell, Democratic senator from Rhode Island, 1961-1997, cares deeply about the family name. When I was just out of college in 1959 and still single, he reminded me that Pells have a tradition of distant cousins marrying each other. In those pre-feminist days, a wife always took the husband's surname. "If you married your cousin Robert," he suggested brightly, "You wouldn't ever have to lose the name."

As his remark showed, mine is a family in love with itself. People often want to know who their ancestors were, and my relatives over the years have found out a lot. In looking for my roots, I have had an astonishing amount of material to work from, including a privately published historical journal written by relatives and called—I'm not kidding—*Pelliana*. There are also archives stored at Fort Ticonderoga, the strategic 18th-century fort between Lake Champlain and Lake George in upstate New York, which was bought by William Ferris Pell in 1820; it is one of the few privately-owned national monuments in the U.S.

When I was in third grade, I could tell the class that our scholarly English forefather, Dr. John Pell, invented the division sign in 1659. His younger brother, Thomas, was a military medic and the one who bought all that land from the Indians. Because Thomas Pell had helped New York's colonial governors, the Duke of York conferred upon him the "Lordshipp and Manner" of Pelham—confirming his title to the property along with various feudal rights, privileges, and obligations. According to Claiborne, this makes us unique, since we are the only colonial family whose land-holdings stem from grants from both the Indians and the Crown.

A paragraph from an 1872 book, *The Olden Time in New York*, caught the spirit of the manor lords and their descendants. It describes how the colonial families had reproduced the English feudal system, complete with a rigid social hierarchy. The top families intermarried, took the choice government positions, and blithely assumed the prerogatives of a ruling class. The author puts

Reenactment of Treaty passing from Chief Wampage to Thomas Pell

it nicely: "They were the gentry of the country, to whom the country, without a rebellious thought, took off its hat."

Even now, in the 21st century, some of my family still behaves like gentry to whom the country should take off its hat. Newspaper writers call us "the ancestral Pells," who have always been "in society." My cousin Toby, son of the senator, is a tall, handsome, gray-haired man with a wide smile and dashing white forelock. He used to head the Newport Preservation Society, a multi-million-dollar complex of museums and colonial homes, whose main office happened to be in a mansion on Bellevue Avenue once owned by his grandfather, a residence that, because of its tile roof, was referred to as "Taco Pell."

Like Toby, I grew up assuming that my family was special. But once I grew older and learned how odd those attitudes were, how contrary to the American democratic ideal, I developed a fascination with the family's remarkable sense of entitlement. Plainly, what I thought was normal was anything but. I wondered if there was a connection between the superiority we felt to ordinary people and the remarkable number of divorces my parents and their siblings ran up; if the sharp-tongued, acerbic style of my relatives (and all the trouble they got into) stemmed from their sense of privilege. Did "sense of privilege"—and its failure—contribute to the suicide of my brother, who shot himself at 26?

I began thinking about these issues in my thirties, when I was married and living in San Francisco with my then husband and three small sons. For many years, I took notes when I went East to visit my parents, writing down the stories they told, watching them interact with servants and equals, then squirreling the notes away, feeling disloyal and fearful of being discovered.

I went through "some changes," as we called them, in the sixties and seventies as feminism and the civil rights movement shook up the attitudes I had inherited. I stopped teaching in a girls' school and became an investigative reporter, deeply involved with prison reform—and with the Black Panthers and the Soledad Brothers. I divorced once, and then again.

My new mindset appalled my father, whose passion was good form, especially when playing the obscure British sports like court tennis and racquets in the private clubs that were so dear to him. He was particularly enraged by an article I wrote for *Women's Sports* about men's clubs that discriminate against women, and he became terrified upon hearing rumors that I was preparing to write a book about the family. In fact, he stopped speaking to me. He tried to frighten me into submission with the threat of disinheritance. We remained estranged for many years, and, though we had a reconciliation a few years before he died, he wrote me out of his will anyway.

In 1688, the Huguenots, Protestants who fled France to escape religious persecution, bought the land which is now New Rochelle. They paid a goodly sum to my ancestor, Sir John Pell, second Lord of the Manor of Pelham. Besides the money, the deed called for the payment of one "fatt calfe every fouer and twentyeth day of June yearly & Every Year forever, if demanded" by his heirs.

In the colonial era, the presentation of the "fatt calfe" was a holiday marked with feasting and fun. Then, for nearly two hundred years, the Pells did not ask for one. But 60 years ago, a distant cousin, Walden Pell, an Episcopal minister who had

At the Fatt Calfe Dinner, New Rochelle, 1956: Sir John, in the painting; Eve's father, Clarry, in white jacket; to his left, seated, Claiborne; to his left, Eve; behind her, with glasses, Rodman.

founded St. Andrew's School, near Wilmington, Delaware, reminded city leaders of the old deed and demanded the tribute. Amazingly enough, the city paid up with a calf. And, irregularly since then, the demand has been made and fulfilled with a range of ceremonies that have featured colonial costumes, flags, speeches, and, occasionally, a local beauty queen.

Once, a New Rochelle mayor refused to honor the contract. Tongue in cheek, the family responded. There were skilled Wall Street lawyers in the family, a Pell spokesman said, who would take legal action to get our city back if a calf was not forthcoming. Instead of seizing farmers' cottages and plows, the family would claim tennis courts and three-car garages. The local press revelled in this historical squabble, and, to the immense satisfaction of my family, the city backed down. Since then, New Rochelle has come up with a "fatt calfe" whenever asked to do so.

I typically postpone my ironing until there's nothing wearable left, so it was not surprising one summer afternoon a few years ago, that my wardrobe hit the wall. Every decent article of clothing I owned lay crumpled and forlorn in a plastic basket. If I ever wanted to leave the house again, there was nothing to do but to rummage in the upstairs closet behind the suitcases for the ironing board, flip on country music for distraction, and get to work.

As the iron heated up, I began sprinkling dots of water onto a denim skirt. I was humming along with a song on KYCY Young Country, when the d.j. launched into a little quiz. "Four tickets to the Mighty Morphin Power Rangers in San Jose," she said, "to the first listener with the right answer."

"Listen up, folks!" she commanded. "Griswold Lorillard was the first man in the United States to do something. What was it?" Astonished, I dropped a big blob of water onto the skirt. Young Country asking about Griswold Lorillard?

She began making fun of the name—"What kind of guy is called Grizz-wold Lorr-ill-ard anyway?" she asked, with a laugh. Then she gave the number to call.

I knew the answer.

Griswold Lorillard was a long dead cousin. I had been researching the lives of my great-grandparents and grandparents, in particular those who lived at Tuxedo Park, the country retreat for Manhattan's old Knickerbocker descendants and its more refined robber barons. From the Gilded Age to World War II, these elite families inhabited their Tuxedo mansions for a few weeks in spring and fall, passing the time between the winter season in New York City and the summer season in Newport.

In 1885, my great-great-great grandfather, tobacco magnate Pierre Lorillard, sold his Newport mansion, The Breakers, to Cornelius Vanderbilt II, and, on a whim, plunged the proceeds into creating a new "colony," as such resorts were then called.

He acquired 7,000 acres of wilderness just 40 miles northwest of New York City and immediately had them fenced in. His overseers hired 1,800 Slovak and Italian laborers straight from Ellis Island and set them to work on a luxurious clubhouse, roads, and a sewer system, all to be completed in just eight months. He ordered up 40-room "cottages," it was said, the way other men ordered a brandy and soda.

Newspapers reported that "Prince Pierre," as he was known, prided himself on being America's biggest spender. He had the best horses, the best carriages, the best cooks, wines, and cigars. At his entertainments, footmen served champagne from great glass pitchers, with duck and terrapin specially raised for his tables. Iroquois, his great racehorse, was the first American horse to win the English Derby.

Griswold was his son. On a grand tour of Europe while a young man, as sthe story goes, Griswold attended a party for the Prince of Wales. Standard evening dress at the time was the tail coat. But Grizzy admired the short jacket he saw the Prince of Wales wearing, and he wore this innovation — with a scarlet waistcoat — at the first Tuxedo Autumn Ball. He was ridiculed for looking like a "royal footman," but the "dinner" jacket caught on and took the name of the place where it was first worn. (An update is in order here: when I looked at the Tuxedo Park website recently, I saw that Grizzy's claim to fame may be more legend than fact. But it was good enough for Young Country KYCY.)

My paternal great-grandparents, Herbert Claiborne Pell and his wife, Katherine Lorillard Kernochan Pell, were among the first members of the Tuxedo Park colony. My maternal great-grandparents, Tilfords and Mortimers, followed soon after.

When I last visited Tuxedo, in 1996, I stayed with cousins who had lived there for decades. Returning to their house after a day's work in the little town library, I showed them a clipping from the forties about a man named Nathan Berkman, who bought a lakeshore mansion from Angier Biddle Duke. As a Jew, Berkman could not join the Club and therefore he was also not permitted to use the lake. "He has been specifically forbidden to jump into, boat on, or dip a fish-hook into the precious pond," wrote a New York columnist at the time. Moreover, when Berkman held a wedding reception at his house and carefully placed signs on the Park's confusing road system to direct his guests, someone turned the signs around and jumbled them so that the guests got hopelessly lost and the party was spoiled. "Oh, I was one of the kids who did that," said my cousin's husband with a nostalgic smile. I bit my tongue and said nothing. I didn't want to argue with him and have to leave their house.

These days, though my cousins' attitude may not have changed much, times are very different. As he and I drove by the Club one Saturday, he looked at the rows of cars parked outside. "You know, lots of the Club's income these days comes from bar mitzvahs," he said. "They must be having one today. Families from New Jersey rent the place." Though resigned to this state of affairs, he still kept count of how many Jewish members the Club has taken in. In 1995, the first black member, a New York Knick named Greg Anthony, had been accepted. My cousin was still distressed. "I knew this was coming," she said, "But I think it was too soon." (Whoopi Goldberg now has not one, but two houses inside the gates…and my cousins have moved away.)

My other maternal great-grandfather, Henry Morgan Tilford, was born in Kentucky and went to New York, where he became a high executive of John D. Rockefeller's Standard Oil Company. According to my Aunt Goody, after Teddy Roosevelt's anti-trust laws broke the oil giant up into regional companies, Tilford was named president of Standard Oil of California—even though he had never ventured Out West. She told me that company officials suggested that he take Mrs. Tilford with him in a private railroad car across the country in order to visit the offices and staff of the newly formed company in the Golden State. But, she continued, rather than endure a journey to what he viewed as a very uncivilized region, he resigned on the spot. By then, it would seem, Mr. Tilford was far too much of a New York gentleman to sully his shoes with California soil.

Eve, 17, at one of her coming out parties in 1954, with her mother, 36

(My mother inherited some of this attitude: when I was first pregnant and living in San Francisco, she assumed I would come East for the birth. "You're not going to have a baby in California, are you?" she asked, as though I might be in outer Mongolia.)

I'd always enjoyed the stories I heard about Great-grandpa Tilford, and since the executive offices of the old Standard Oil of California (now called Chevron) were in San Francisco, I went to the company's History Room to find out more details about his business experiences.

I was shocked to discover that he had been every bit as energetic and resourceful as the Mortimers had been foolish and lazy. When still very young, he and two of his brothers left Kentucky for the frontier and went into the oil business. Contrary to what the family thought, they spent years in California knocking heads with competitors, gouging advantageous contracts with railroads, and scrambling for market share. They were smart, aggressive, and enterprising, devising new and profitable ways to refine and transport kerosene...until Standard Oil bought them out and made them executives with offices in New York.

He continued to succeed. In company histories, Henry M. Tilford stands out as the man who, against all existing wisdom, authorized the huge gamble of wildcatting for oil on the West Coast. It paid off fabulously.

Armed with this information, I went to tea with Aunt Goody on my next trip to New York and told her what I had found. To my surprise, she refused to believe a word of it. "He never went to California," she said crossly. She stuck to the family version, that he had resigned on the spot in Manhattan rather than visit such an uncouth place. Certainly, he never worked there.

After tea, as I was leaving and putting on my coat, Goody grumbled, "Wasn't it fresh of Standard Oil to make up all that!"

Stanley Mortimer, my Grandpa who later lost all his money, was a good athlete. In one alcove on the far side of the Big Room at his Tuxedo Park house stood mahogany cases for his silver trophies, shelves and shelves of them, large and small. After they were stolen in a celebrated burglary, my grandmother, whom we called Gargy, had them all replaced in sterling silver—whether the originals had been sterling or not.

Grandpa had a few other interests, among them Tuxedo's volunteer fire department of which he was a ranking member. One night in 1916, he was disturbed at dinner by the shriek of the fire alarm. Rushing to the firehouse, he and the other gentlemen hitched up the hook-and-ladder and, clinging to the apparatus with one hand, changed from their dinner jackets into red flannel shirts, rubber boots, and fire helmets as they sped to a burning inn a few miles away. They arrived only to discover that they had no source of water and no buckets for a bucket brigade.

Their wives followed. "The flames illumined the faces of many society women in the automobiles parked in a circle about the hotel—women wearing heavy furs over the evening gowns in which they had abandoned dinner tables," a *New York Herald* reporter wrote. The luckless inn burned to the ground.

Afterward, the Tuxedo Park residents returned to their dinner

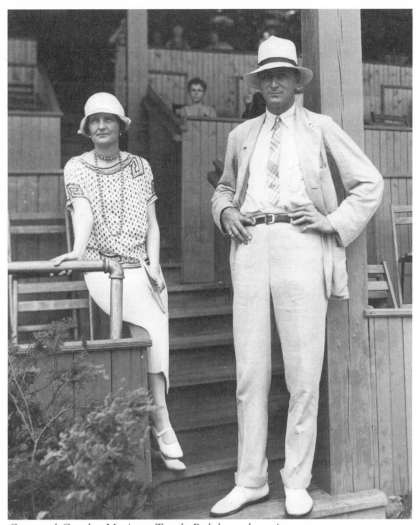

Gargy and Grandpa Mortimer, Tuxedo Park horse show, circa 1930

tables, the men being permitted to resume their places still wearing their red flannel shirts and high rubber boots, while the butlers held their fire helmets.

Because of Gargy's Tilford money, it didn't matter from a financial point of view that Grandpa had no real job. Without an office, however, he needed a place to go during the day. Wherever he was—Tuxedo, New York, or Palm Beach—he belonged to a club. He arrived at ten and picked up the paper. The joke was that by the time his friends arrived for lunch, he had read as far as page two. Gargy referred to his club as his "sandbox," as if he were a child. Once, long after Grandpa had died and a male houseguest wanted her attention, she complained with great frustration to my Aunt Goody, "What do you do with a man who won't go to his club?"

During my childhood, I particularly liked the splendor of Gargy's Tuxedo house, Keewaydin, which was once owned by Pierre Lorillard. Designed by the renowned architect Stanford White on the highest point in the Park, it was far grander than the places my mother or father owned. A stone-and-brown-shingle villa with two enormous round towers, the house was dominated by a long, splendid living room that you stepped down into from a double stairway leading off the capacious front hall. Whenever I descended those steps, I felt like a princess making an entrance. I loved to lie on the huge polar bear rug before the massive stone fireplace, my face nestled in the white fur. The darkly paneled room was so big that the various clusters of comfortable sofas and chairs seemed miles apart, like separate little villages.

I was never to bother Gargy, whose prayer times in the morning and nap times in the afternoon were especially inviolate. She didn't make pies with me; in fact, she never entered her own kitchen. She didn't brush my hair, or comfort me if things went wrong, or touch me, except for little air kisses.

Gargy's servants addressed me as "Miss Topsy," which was my father's nickname for me, after the "pickaninny" in *Uncle Tom's Cabin,* since my complexion was dark and I had curly dark hair. I stayed on the third floor with my young uncles, John Jay and Dickie, my mother's much younger brothers, who were not far from my age. Gargy had had four children, including my mother and Aunt Goody; then, after a long lapse, she had the two youngest boys. John Jay, two years older than I, was a baby at my mother's wedding. He was named for our ancestor John Jay, first Chief Justice of the new United States. Miss Grey, the boys' governess, looked after me, too, when I visited. We ate supper early, at the children's table in an alcove of the large dining room, never with my grandparents. We were brought in to see them at tea time in the afternoon.

Down a garden path stood a playhouse equipped with masses of toys, a stove that worked, and small, wheeled wooden horses covered in real hide. Inside the big house, we could ride up and down in the "lift," as Gargy called it—an old-fashioned, tiny, elevator. John Jay introduced me to radio serials, and we would sit on his bed in our pajamas and bathrobes, listening avidly. The *William Tell Overture* to this day takes me back to the third floor at Tuxedo, with its splendid view over the Park, while the adventures of the Lone Ranger and his great white horse, Silver, emanated from the radio.

Gargy saw her own children only when they were clean and polished and at her convenience—hardly an atmosphere in which love, or nurturing, or parenting, could flourish. Real emotion, like earning money, was embarrassing; love was a word more commonly used when playing tennis or in conversation—generally about an inanimate object—as in "I just love this new dress." Not for nothing are we WASPs called "God's frozen people."

It took me a long time to acknowledge the bleakness of these family relationships, a coldness of heart that was masked with jokes, manners, and snobbery. In fact, the denial of feelings seemed indispensable to the glittering life within the Park. The insouciance, the casual acceptance of one's class superiority, could never have survived in an atmosphere of genuine sharing and caring. □

Eve Pell, shown above, age 8, on Grey Bud, at the Piping Rock Club, Long Island, lives in Mill Valley. This is an excerpt from a memoir-in-progress. Born in 1937, educated at Garrison Forest and Bryn Mawr, she was once ranked first among American women distance runners over 60. She continues to work as an investigative journalist and has two grandchildren. E-mail: evepell@aol.com

LONG DAYS MAKE SECRETS ESCAPE

Tsering Wangmo Dhompa

*M*y mother had wanted to introduce Tibet and her family to me. She had been the only one in her family to escape to India in 1959. She had seen many people killed along the way, many die of fatigue or sickness, and she said she wanted to return, because she didn't want to be separated from those she loved. She was not prepared for a life in exile. She was only 20 when she entered Nepal with her husband. The fate of her family left behind haunted her. We planned a trip to Tibet for June, 1994, but she died in a car accident that January 1. I went to Tibet that summer for the first time.

Now, for the second time, I am on my way to my mother's nomadic home in Nangchen, in the eastern province of Tibet. I stop long enough in Lhasa to light butter lamps and make offerings at the sacred sites: the Potala, the Jokhang, Sera and Drepung monasteries, and the small shrines tucked away all over the city. It would be unacceptable for any Tibetan to leave Lhasa without making these gestures. I am not a knowledgeable Buddhist, but I have been brought up to regard sacred places as sacred. Besides, it seems an apposite ritual considering the long journey ahead.

A road trip to Nangchen takes four days and three nights; the fastest way is to fly into Xining. My cousin-brother Abo, and his mother, Tayang, who is my late uncle's wife, are waiting to receive me in Xining. I find out from Tayang that Xining is considered the very essence of modernity and reasonable altitude. Many Nangchen nomads tell me they don't quite have a vocabulary for Beijing's tall buildings and busy subways, but they are comfortable with Xining's big-city indulgences, because they maintain a small-town familiarity. Xining, they say, is big enough to allow for anonymity and small enough to bump into nomads when you want to. In fact, increasing number of Tibetans from the towns of Cherku, Nangchen, and Zatou are choosing to spend more time there after their retirement. They buy apartments in Xining and say the milder winters and lower altitude agrees with them. This is so, especially so for those who suffer from hypertension and high blood pressure, and there are many who do.

We stay in Xining for two days so we can visit Kumbum monastery, which used to be the fifth largest in Tibet. Chinese from the mainland are encouraged to travel into Tibet, and Kumbum is one of their destinations.

We arrive with a busload of Chinese tourists. They wear bright caps and large cameras. The guides, in loud voices, explain the history of the monastery. They also explain the Tibetan people, their Buddhism, and their habits.

While we are making sure that all the members of our party are present, a woman comes up to ask if we need a guide. She is dressed in a blue suit, her skirt opening a little over her calves to enable her to walk in quick short steps.

An elder tells her we do not need the Chinese to tell us our own history. She seems confused for a moment and then walks away without revealing how the words have affected her. Her hair is knotted into a tight bun. It hugs the base of her neck. Her neck is very white and slender, and she holds it stiffly as she walks away from us. I see her smiling at another group as she leads them through the large metal doors into the monastery, where a monk greets them in Chinese.

For a moment, we enjoy a little victory. Small victories are all we can hope for and yet this feeling, too, is ungrounded.

There are more Chinese tourists this day than Tibetan pilgrims. We must accept this joint venture between tourism and spirituality. The world here has been upset for a long time, and the people I am with are used to losing their way. I follow them. My little nephew prostrates himself at the altar and recites a prayer. We are led to a shrine hall with statues of Buddha and Chenresig, the deity of compassion. We spend the day offering butter lamps and praying to the gilded faces, praying for all the beings who are alive and those who have left us but are in their new bodies.

Early next morning, we leave for Cherku, the capital of Yushu Prefecture and home to most of my cousins. It will take about 20 hours by car to get there. The high grasslands go on and on in hypnotizing uniformity. We drive for four hours without seeing a single living person.

At 15,000 feet, miles and miles from Lhasa, the insistent whistle of wind is heard passing through land that has not seen grass for six months. It is June, the snow has melted, but the ground is still cold, and I think it must be a painful process, this period of thawing, even for stones and mud. Soon the grass will grow tall and inhibit easy walking, and then it will be impossible to think of the grasslands as being any other color but green. And yak and sheep will be unable to separate their dreams from their waking wanderings, thinking of summer as heaven's vast feast.

The plains in Golog, Amdo, and Kham are wild and terrifying to those not used to open spaces. The Chinese must think them strange places for habitation, even impossible, with the unending monotony of land and the sky in its abandoned stretches. Yet they are everywhere—the Chinese. They have restaurants, they have billiard rooms, and they have stores where even cattle would beg not to be kept.

The uncommonly expansive beauty of the landscape seems incapable of deception, but under its calm deportment are winds so brutal that yaks are known to die grazing, while their jaws are in chewing bliss. Then there are the sudden changes in the sky—like a genius who cannot help but be offensively whimsical—or the abrupt downpour of hail or rain. Suddenly here, suddenly gone. Sometimes mild and sometimes brutish.

Every morning, I sit out on the verandah of Abu's house and have my tea. From where I sit, the Cherku monastery looks like a party of horses.

Tayang and I sit in the sun facing the hills. Tayang always creates a pile of watermelon seeds at her feet. She lays a seed between her teeth, cracks it right in the center without bruising the seed, and spits out the shell. Seed spittle dangles on her black dress and attempts a precarious dance from her sleeves.

These seeds are scattered all over town. Garbage does not have a designated place. The invention of cans, plastic, and paper was not kind to this town, either, and they are everywhere. Even the river passing through town is garbage gray.

I ask an eleven-year-old girl if she notices anything absent in the water. She says "fish." She has never seen this river look blue. She drops a candy wrapper into the water. A man cycles past us and spits into the river. His spit flies in a perfect arch, almost stops on someone's dress, then plunges into the river. For a moment, I see its entry into the water and then it is carried deeper into the center of the city.

My mother, Tsering Choden Dhompa, was born in Dhompa, a nomadic region best known for its ordinariness. For lack of anything substantial to boast of, we say we are good and loyal people. Or that our dialect is closest to the Tibetan script. In the old days, Nangchen had a king. He presided over 25 different nomadic communities overlooked by chieftains who reported to him. There were four ruling chieftains and of them, the Dhompa chief was well recognized and known. My grandfather was the Dhompa chief.

The road to Dhompa was built recently, and our journey does not take very long, because of the dry weather. Tayang, my twelve-year-old nephew Kunga, and I leave from Nangchen, where we have spent a night with a cousin. The road is hewn into mountain after mountain and we are jostled and shaken all the way. Twice we are stopped on the road by nomads who recognize us — distant cousins — and we take a tea break in their tents. The road ends near the monastery of Bagsigon, but we force our way through shallow streams and walk when the road is treacherously narrow, to arrive exhausted and bruised at the bottom of the hill where my mother's family once lived. The jeep can push no farther, and we walk up the hill to the old home, Tingsikha.

Tingsikha hugs a spectacular view. Because of its height, nothing obstructs the vista of fields, river, and mountains in front of it. The family's old home had been razed to the ground by the Chinese, but Tayang has partially rebuilt it to house a large shrine room, a kitchen, and two bedrooms. The rooms are spartan, and the floor is made of mud. Two pillars hold the roof up. Very little light enters the rooms even during the day, because the rooms have just one small window that stays hidden from the sun. The window is covered by a plastic sheet.

There are five of us — Tamdin and Wori, who are caretakers of the house, Tayang, Kunga, and me — and we immediately adopt an easy routine. Wori looks after Tingsikha, and, except for Tayang and

Tamdin, who spend a few months there every summer, he is alone. Tamdin is an old family friend, married to my mother's cousin-sister. His father was an *Ilonpo*, a local official, and worked closely with my grandfather.

Wori's face is darkened by soot. His eyes are molten and puppy-like. His neck is soot, and, when I tell him to wash his face, he laughs at me in a manner that makes me feel a little foolish, though that is not his intention. Wori resembles the charcoal pillars in the kitchen. He has not bathed for a long, long time, nor has he washed his face. He cannot say for how long, but guesses it could have been a few years. He offers a generous time frame, "some time back," when I ask him. He has never brushed his teeth. He hasn't had to, since he lost most of them when he was 30.

Wori's frame is a mystery. Like most nomads, he has far too many layers of clothes on; rarely is an ankle or elbow exposed. He wears two layers of pants, a sweater, and a *chuba* above that. He is not aware of being hot, not even when beads of sweat gather around his forehead. He wipes them off with the sleeves of his chuba and says the sweat is from the weight of the water bags he carries on his back several times a day. He stands all day and has to be reminded to sit down. He tells me the habit comes from years of tending yaks on the mountains. His hands are often folded over each other, another habit cultivated from his herding years, when he would put one hand over the other alternately to keep them warm.

Wori says he feels most comfortable standing upright, and I catch him fast asleep one evening leaning against the wall. It seemed as though it was the natural way, maybe the only way to sleep. His smile is quiet, like all else about him. It is as though he does not wish to acknowledge his own existence. He simply stands in the shadow of the big stove for hours without knowing he has been there.

Wori has never said no to a Dhompa and has served the family for many years, and I suspect for very little in return. I tell him he ought to retire and let his son look after him. He says yes, steals a look toward Tayang, and stokes the fire. Tayang tells him she needs him. He says yes. I shake my head accusingly at him. A giggle escapes his body.

The kitchen is dark. (The darker and dirtier the kitchen, the more auspicious it is for the family. The cooks at Tingsikha took great pride in the dingy kitchen and in its oil-dank cobwebs, sooty beams, and light-shunning roof. Tayang tells me proudly that it was a dirty, dirty kitchen. Perhaps even the dirtiest!) The cooks were all men in the old days. Women weren't allowed into the kitchen. The reasons are ambiguous, but it has something to do with the deity or the patron of the stove being stifled by the presence of women. Today, some women enter the dark room, but the women who look after the family's yaks and sheep do not. They continue to adhere to the tradition, and no one forces them in. They cannot give reasons for their banishment from the kitchen, but state that they ought not to enter.

Tayang and I brought some vegetables and fruits with us, because we knew once we were in Dhompa we'd have no chance to go to a store and shop for fresh fruit or vegetables. We have a bag of apples, pears, cabbage, and gourd to last us a week. I offer Wori and Yungyang a pear. They are eating one for the first time. Wori wants to know if it is a fruit or a vegetable. He is quite sure it is a turnip. Should he boil it? Yungyang, who once was responsible for milking the female yaks owned by my grandparents and who is visiting us for a week, needs a knife to skin her pear. She has decided to eat it, not save it for her grandson. Everything else I share with her goes into a plastic bag she keeps in the corner of the kitchen: six sticks of chewing gum, squares of chocolate stuck to the wrapper, and hard candy. She puts the pear on the sheepskin she is kneading, then wipes her hand with ash. Her pear is covered with ash and yak fur. It has become a giant kiwi.

"The Chinese," she says, "are clever to make such tasty fruit."

She sinks her three good teeth into the fruit. Her pupils dilate as her tongue registers the sensation of pear.

"It is like, like…" she says, not sure what to compare it to.

"It is like a fruit," she lets out.

She wears a deep brown shirt. It is a handsome color on her and handsomely coated with dirt. She has rows and rows of *sungdu,* blessed sacred threads, around her neck for protection. They symbolize security and faith. Like many Tibetans, she also wears prayer shells around her neck, serving as ornament, but also as a mark of achievement in having said a million prayers. The shells are from her prayer wheel. She studies my sungdu and a locket that contains hair and nails of expired lamas. Precious relics my mother put together so I will keep safe.

There is not too much room for innovative cooking in Tingsikha. We have meat, yogurt, milk, tsampa, plus the barley flour, salt, white flour, and the vegetables we brought with us. Wori has spinach growing in the garden. I am the designated cook and try to do the best I can. Not only am I hampered by the lack of ingredients, but my friends are used to a certain diet and they don't have much tolerance for change. I stick to noodles every night but try variations. I make *thenthuk,* long noodles, *bhakhthuk,* round flour balls shaped like miniature conch shells, and *mothuk,* dumpling soup. Wori, Tamdin, and Yungyang have not had bhakhthuk for a long time, and they marvel that I can make food from "the old days."

"Oh, we have forgotten the old ways," Yungyang says. Not one of them knows how to make bhakhthuk, and I like the irony that I am the one to make it for them. Tamdin is the only other person who knows how to cook in our group, but he says he is no good, so he restricts his role to kneading the flour. To make bhakhthuk, I make long thin ropes with the soft kneaded flour, then take small balls in my hand, dig my thumb in it against my palm, and drop it in the pot of boiling broth. Wori and Yungyang don't cook at all. Nomads are not adept with condiments. They either boil meat, eat tsampa with tea, or cook *thukpa.* When they make thukpa, it is not much different from boiling meat. They simply add chunks of meat and boil them together. Taste and delicacy are not a priority.

We are good eaters. We waste nothing. Extra milk is made into butter or yogurt. Curdled milk is made into dry cheese. When the yogurt gets too sour for us, Tamdin makes *churtsa,* which is a sour tsampa soup. He stirs the yogurt with some water and brings it to a boil. Yogurt smell takes over the room. He adds dry cheese to it with some salt, meat, and tsampa and keeps stirring the mixture till the broth is thick and creamy. It is wonderful after a long or a cold day. Each mouthful a sour benediction. On long summer days, we pray for yogurt to sour.

The apples begin to turn soft and, fearing their ruination, I proceed to make jam. I have never made jam and don't know how to, but I forge ahead. I cook the apples till they soften. Then I slice them in tiny pieces and cook them in butter till they dissolve into a creamy paste. It does not taste like jam, but I am not worried because Wori and Yungyang have never tasted jam and probably never will. I apply the paste on a slice of bread and offer it to them. Wori smiles and tells me he thinks I am a genius. Yungyang is not sure she likes it.

For fuel we use yak dung, collected into little mounds and plastered on walls. When dry, these are torn off the walls and stored in the kitchen. Some prefer to spread them in thin paste on the ground, which they simply tear off when hard. Wori chooses to build a dung stupa on the grassy field outside the house. A good year is measured, he tells me, by the height of the mounds of yak dung. Our yak-dung stupa is 20 feet high.

Kunga tells me he is not taught about Tibet's takeover in his school. He gives me an example of his history teacher who introduced the Nangchen King in one sentence: "The Nangchen King was a bad man who thought only of himself."

Kunga and his classmates took the teacher into the bathroom and beat him up soon after the lesson. The teacher was Chinese. Most of the students were from Nangchen or had friends from Nangchen. He thinks it wasn't a political message against the Chinese government, they simply didn't like the teacher and his insensitivity to their being from Nangchen.

I tell him about the hundred thousand Tibetans who escaped to India in 1959, who are now all over the world, and the million Tibetans who were killed since the Chinese takeover. We talk about our grandparents and many other relatives who died in prisons and in labor camps. There are gaps between his sense of history and my sense of history.

His Tibet begins after 1959, because he was born here. He cannot imagine living anywhere else but in Tibet. He thinks of himself as a Tibetan, but does not think he could live as a herder. He likes wide streets, electricity, and cars. He sees Dhompa as a place that needs to develop. It is not yet a livable place. Dhompa is pretty, he says, but he misses his friends and doesn't know how to keep himself entertained among hills and cattle. I've lived in so many places I don't quite know where home is, but, like him, there is no doubt that I identify myself as a Tibetan, even though I have no piece of paper that gives me the authority to do so.

The Chinese government is building a road below Tingsikha

that will link Dhompa to other nomadic regions. There are a thousand workers brought in to work on this road, and they are all Chinese Muslims. Their white caps dot the green fields like white daisies, and their cream tents are neat and orderly. They have come from places that are warmer, where they do not eat yak or wear wool all year round. They have light and very smooth skin. Even after a full day in the sun, they look fresh and youthful. They cannot look as though they belong to Dhompa even if they want to. They catch fish in the Kichu by throwing dynamite into the water. They cannot believe the Tibetans don't fish when the river is rich with free food. Until the Chinese arrived in Dhompa, nobody had ever been seen catching fish. Big animals were eaten. The calculation was simple: one yak fed a family for months. One fish fed one person for one meal. Life was counted in numbers. Outside thei workers' tents are not yaks and sheep, but piles of provision for the coming months: small tractors, shovels, metal rods and implements of various shapes, and cans of food.

The overseer of the road project happens to be a man from Nangchen and a distant relative—an uncle of a cousin's wife's husband's cousin's wife—so I am sent to invite him to our house whenever he needs anything. He is sitting outside his tent drinking beer when I visit him with fresh milk and yogurt. His men are skinning a yak they have purchased from a herder. (The skin comes off easily; the task looks effortless.) He thanks me for the kettle of milk without glancing at it and says he will eat the yogurt after dinner. He then instructs the men to cut meat from the thighs and shoulders for me to take for Tayang.

It is ten in the morning, and his face and eyes are red and swollen. He yells for his cook and, on spying him, tells him to cook a good lunch for me. He tells me the Chinese Muslims make good workers, as though we are in mid-conversation. They can stay away from home and do a good job. They have a special gift for working on land. "Tibetans," he says and rolls his head. His red eyes bulge out. "Tibetans like picnics. They must eat several times a day, they must rest several times a day. For them, every working day is a picnic. Utterly lazy."

He asks me if I was witness to the war between the moon and the demon the previous month. I tell him I was aware of the lunar eclipse, but had gone to sleep quite early. The beer pushes his words out in a loud grunt.

"What is the reason for the moon sleeping at night?"

"Well, according to science…" He doesn't let me finish.

"Oh, you are like the Chinese. Do you know that the real reason is that the moon was being attacked by the demon. The moon was defenseless but for its virtues."

He says he played an important role in going to the aid of the moon. He fired guns into the air. Twenty times or more. He also got his men to bang tin plates with a spoon.

"Great racket we made," he says.

He looks so proud of himself. He opens another beer. I tell him he should eat instead of drinking beer. He complains of his heart problem and hypertension and takes a swig of his beer.

"And then I blasted four dynamites. You should have heard the sound. It was the dynamite that brought back the blush in the moon."

His face is the color of burnt milk spilled along the sides of a bowl. He could easily be mistaken for a child's idea of the evil sorcerer. He tells me about my mother, how wonderful she was, how wise she was. I agree with everything he says.

It is mid-July in Dhompa. We have six scorching days and no rain. Yungyang says it's because the Chinese Muslims' road workers prayed to their god for dry days so they could finish their job quickly and return home. She says she saw them kneel in the sun for a long time. Of course, no one is there to verify that they were really asking for dry weather, but everyone at Tingsikha has accepted this answer.

"What about the dynamite?" I ask Yungyang.

The dynamite and the Muslim gods, she says, are working together on this.

It rains on the seventh day. We run into the kitchen. We have milk, we have yogurt, and our mud stove is burning all day. Our garden has spinach. Our cattle are eating grass and wild flowers, now strengthened by rain. The river has water. Our roof keeps out rain. Our world is small and simple.

Long days make secrets escape, and the stories are made sweeter because the listeners are easily entertained. It is revealed that Yungyang was the one with a whistle during the Cultural Revolution. She would blow the whistle to alert Tibetans to attend the re-education meetings in the mornings and evenings. This was an enviable job, because it spared her the punishment others received. She witnessed their singing and their beatings. Those who were there remember how sharp and serious she was about her whistle-blowing duty. She denies it was ever her, but Tamdin says it is unlikely he could forget the face of someone he knew who stood in front of him day after day. Those days of terror, now recalled over tea and dried meat, become occasions of comic interlude.

They sing the songs they were taught during the Cultural Revolution. Their voices quiver, their pronunciation so Tibetan.

"Long live Mao," they sing. "A thousand years to Mao."

I tease them that their Chinese hasn't improved in all these years and realize it must have been such a struggle to learn the words to the songs. They say they suffered so much just trying to memorize the lines. It was unreasonable, they say, to expect people to remember words in a language they didn't know. But that was the way it was. So many beatings, Tayang says, because their silly heads wouldn't keep the words in order.

They sang to Mao every morning and evening and were punished if they stumbled over the words. They were not allowed to say their prayers. Mao was the focus of their devotion, and they sang to him in his language. "Some people visualized a deity instead of the picture of Mao, so they sang willingly to their deities," explains Tamdin.

The words are warbled, but they emerge out of the years of hibernation. It's almost another language.

The mind is a powerful place and cannot be imprisoned, I am told. Even in the midst of such madness—and surely it was madness that made the guards inflict so much physical and mental suffering—they were able to endure without becoming broken, bitter, and sad people.

On such long warm days out on the grass, their passivity and congeniality upset me. It does not appear that these are people who have suffered and watched families disappear before them. Tamdin's parents were killed in prison—he says this as he stitches a coat without any display of emotion. He was in prison for a few years, but doesn't talk about it. He wears his suffering with the same acceptance as he wears his old clothes. The world around him is the way it is, and he doesn't have any desire to make changes.

They would prefer if the Chinese left them alone, but since

that is unlikely, they continue with life. Even the older people move from a sense of loss to a state of repose, and in this repose they see continuity. The loss was due to their karma and therefore the Chinese are part of that karma. That is their belief.

And like them, I resort to nothing.

Time nibbles into summer months and, without our knowing, it is time to leave Dhompa and return to Cherku. The air cools, the grass browns. It is almost August. In a month Tingsikha will be cold and dreary. Wori is already preparing for the months when he will be alone. He dries grass and the leaves of the radish plant on the eaves and roof of the house, airs his winter clothes, and gathers as much dung as he can. He piles the dried dung in the kitchen. He has butter stored in leather pouches, semi-dry yak meat, some dry meat, dried cheese, and bags of tsampa. He has a lot of work the next two months in preparation for the cold months. The families who live behind Tingsikha will be back, too, so he will have neighbors.

When it is time for us to leave, he stands back against the kitchen wall. Our goodbye is hasty. Tibetans are not often sentimental, and our emotions don't often find their place in words. He says maybe we will not see each other again and waves us off to where the road begins. □

Tsering Wangmo Dhompa grew up in Nepal and India. She works for the American Himalayan Foundation in San Francisco. Her second collection of poems, In The Absent Everyday, *was published this winter by Apogee Press, Berkeley. E-mail: dhompa@yahoo.com*

FROM THE WINGS

Peter Wright as told to Alan Gullette

I was eleven when World War II began, and I was immediately evacuated, like the rest of St. Francis Xavier's School in Liverpool, to North Wales. Specifically, I was sent to Saint Asaph, the smallest city in Britain, only nine miles from where my family lived in Llangollen. The whole thing was a bit silly.

We continued to live there—students and masters—for the nine months of the Phony War. Everybody, especially the city kids, were so bored with the countryside, they just wanted to get back.

We arrived back just in time for the Blitz. We were told that if an air-raid siren sounded on our way to school, we had to immediately go to the nearest shelter. In fact, if the raid continued after midday, when it was over, we just went home: school was cancelled. We used to pray for air raids!

The raids during the day were, in fact, just reconnaissance, checking on the damage caused the night before. There were thousand-bomber raids hitting Liverpool, Manchester, Coventry, etc. We were very close to Mt. Snowden, the highest mountain in England and Wales, and we could tell where the aircraft, using Snowden as a guide, were headed at night: if they flew to the left, Liverpool was getting it; if they flew to the right, it was time for Manchester.

Our family was allowed one gallon of petrol a month, and it was used to go to Chester, 20 miles away, to see a movie. On the night we saw *Pinocchio*, we got back to find the Home Guard blocking the entrance to our house. An unexploded bomb had been dropped in our garden! It was probably from a damaged plane jettisoning its load before returning.

It took seven days to defuse the bomb. In the meantime, my mother got together with her best friend and sister-in-law, my aunt Hannah (they were both named Hannah), and said, "I have to get into the house because they are going back to school and they've no underpants or anything."

They stole under the fence and got various things for us to go back to school with, plus my mother's fur coat and jewels. As they were trying to leave, they were arrested!

They appeared in court the next day, before a young judge who was not in the Army due to a physical problem: one leg was shorter than the other. He chastised them and was putting them on two years' probation, when my aunt said, "Bobby Jones, I knew you when you were having your nappies changed. Don't you dare speak to me like that!"

They left the court and, of course, never, ever reported for probation. My father enjoyed it enormously. When he came home on leave, he introduced my mother as, "my wife, a member of the criminal classes!"

Francis Bacon

At 18, I was conscripted into the armed services. Fortunately, during my first three months, I was stationed just outside London. This was especially wonderful because no one who lived in the countryside was allowed to go into the city—it was too dangerous. But every second I could get out of those barracks, I was in London.

One night, I was standing outside a theatre, and a man came up behind me and said, "Hey, soldier. Are you going to see the play?"

I said, "No, I can't afford it."

He said, "Why don't you come with me and see it from the wings?"

And that is where my title comes from. I have been very privileged in my life and have seen so many splendid people and things and happenings "from the wings."

During that first sojourn in London, I wandered in and out of galleries with my eyes wide open. One day, through a gallery window, I saw a painting I thought was wonderful. It was by Francis Bacon. I walked in and started looking at the exhibition.

In those days, nothing seemed shabbier than a private's

uniform. You could be a millionaire, but you looked as though you were on the edge of poverty.

I asked the girl behind the desk, "Excuse me, but how much are these works?"

"They're all sold."

I pointed to one and asked, "How much would that be if it were not sold?"

She said, "Oh, that wasn't for sale. The artist is keeping it for himself."

I turned to her in disgust and said, "Please congratulate the artist: he has chosen the best painting in the exhibition." And I walked out.

All this time, there had been a man sitting reading the newspaper whom I had barely noticed. He followed me outside and called out, "Hey, soldier! Let me take you to lunch."

It was Francis Bacon.

I didn't become a close friend of his, but I used to see him occasionally. After I moved to America, I would visit my parents in Leamington Spa every Christmas. And on the drive up from the airport, I would stop to wish Francis a happy Christmas.

One afternoon, I walked in and he was on the telephone. He told me to get myself a gin and t. I went into the drawing room and there—in the company of a Picasso, a Matisse, and a Léger— was the painting that, over the years, had been called "Peter's painting," the one that was not for sale.

He followed me in, gave me a hug, and said, "Peter, it is time you had your painting. But you have to pay for it."

I said, "Thank you, Francis. It would have been wonderful in the past, but I can't afford the £100,000 you are bringing now!"

He said "Oh, no. I want you to pay for it, but at the original price: 500 guineas."

And so, after all these years, I had "Peter's painting." Unfortunately, much later, my home and business in upstate New York burned to the ground, and that was one of the many works that was destroyed.

Evelyn Waugh

At university, I joined the guild of Saint Francis de Sales, a medieval guild of scribes, modernized for writers and journalists. I was just 21 and felt privileged to go to meetings once a month and drink sherry with the likes of Graham Greene and Evelyn Waugh.

At one meeting, Waugh denounced the National Health Service. He said that looking after health was the job of parents and family. I got up and argued with him. I remember telling my father I told Waugh exactly what I thought. My father said, "You are far too young to challenge a man of that stature." But I did.

I was friendly with his son, Auberon, and whenever my name was mentioned, Evelyn referred to me as "that raving socialist friend of my son's."

After our altercations, I read, just as it came out, Evelyn's novel *Helena* (1950). It is the story of the daughter of Old King Cole, the woman who gave birth to Constantine and discovered the True Cross. It was—and still is—one of the most beautiful books I have ever read. And, although there was a little tension between us, I had to go up and thank him for it.

He looked at me strangely for a moment and then he said, "Oh, yes, Peter. You are so young. It's especially privileged that the young love the book." And he added, "You know, it's the only one of my books that has not had wonderful reviews. And in my mind, it will be the only one I'll be remembered by in a hundred years' time."

He smiled at me and patted me on the shoulder.

Much later, I was having lunch with Graham Greene and he was talking about reading: "When I read, I always want to change words—with one exception: I have never been able to think of an alternative word in any of Evelyn's books. He is so perfect!"

The Queen Mum

When I first went to work at Covent Garden, my immediate boss was Bill Beresford, an Australian who was lazy, which meant he passed most jobs over to me. I didn't mind because, during the course of my first month, I met two of the most famous women in the world: Elizabeth, the Queen Mother, and Maria Callas, the diva.

I had been told that when the Queen Mother appeared at the Opera House, one had to meet her at the door and get her upstairs to the royal box as quickly as possible. She was referred to as "the late Queen." There was no point in the curtain going up until she had taken her place in the royal box, because the whole theatre would get up and clap. No other member of the royal family got this response.

The strange thing about the royal box is that even if you stretch out over the edge as far as you can go, you will still see only half the stage. Which didn't matter much, because few of the royals were much interested in the arts.

Behind the royal box, there is a small sitting room. When the Queen Mother arrived, there would always be a bottle of Teacher's Scotch (open, with a little drop out of it) and a packet of Passing Clouds cigarettes (also open, with one out).

I met the Queen Mother at the door and escorted her upstairs with her lady-in-waiting. I poured out a glass of the Scotch for the Queen Mother.

She said, "Mr. Wright, I've had a lunch today, and I'm tired of these official lunches—they always give us chicken. I don't really like chicken, and if I do eat it, I prefer it to be dark meat, but, of course, they give us breast. I'm so hungry. Do they still have that sandwich bar downstairs?"

I said, "Yes."

She said, "You couldn't possibly…"

"No problem! Any particular—"

"Not chicken."

I went downstairs. The sandwich bar was staffed by two old queens. I told them what I needed, and we decided on a smoked salmon sandwich. They handed it to me quite plainly on a plate. I asked, "Haven't you got some parsley or something to garnish it with?"

"No."

But they did have little boxes of raspberries, so I took one and emptied it around the sandwich.

When I took it up to the Queen Mum, she smiled and said, "Oh, I love smoked salmon—and how clever not to waste the soft-fruit season!"

Maria Callas

I was given the job of driving Maria Callas to Pinewood Studios, where she was going to be interviewed. (At that time, nobody realized what a bad driver I was!)

I felt rather awkward. She just sat there for a while, and then told me that I drove very well. Then she said she hadn't realized how far the studio was. I said we would be there very soon.

She had recently caused quite a stir by ranting and raving at the Vienna Opera House. Then she failed to appear at the reopening of the Rome Opera House after the war, and the president of Italy was left stranded. As a result, there was great criticism of her in the continental press.

I just couldn't help myself and asked her why there was always such tension between her and the opera houses in Europe.

"Because," she said, "the only gentlemen left in Europe are on this side of the Channel."

The Wolfenden Report, John Cranko, Sir John Gielgud

Until the Wolfenden Report came out and the 1967 Sexual Offenses Act was passed, homosexuality was still a crime in England. It had gently changed since the time of Oscar Wilde, who was sent to jail for two years of hard labor. After 1956, if you were caught, you only had to pay a fine of £5 or £25. (One never found out what you had to be doing differently to pay five times as much.)

Just before the law was revised, there was a big effort to catch men having sex with each other, especially in public places. During this time, several prominent people suffered embarrassment and sometimes even the destruction of their careers.

There were various meeting places for men seeking men publicly, usually railway stations—Paddington and Victoria, in particular. Also, the Chelsea Embankment, which every few yards had a toilet; there, local guardsmen, especially, would wait for people to accost them and give them £10 so that they could play with each other.

Several people whom I knew personally and loved dearly were involved in this final attack on the homosexual world.

John Cranko, the South African-born choreographer, had captured the London ballet world, first with his breathtakingly beautiful *Harlequin* in April, and then with his very amusing adaptations of *H.M.S. Pinafore* and *The Pirates of Penzance*. He was arrested on the Chelsea Embankment and called me, in tears, at three a.m. I told him to get into a taxi and come over, and, in the morning, I would go with him to the Bow Street Court. I said he should plead guilty, but tell the magistrate he was an accountant or a clerk.

This he did. The only problem was that John had a Cyrano

de Bergerac nose. Once you had known him for five minutes, you didn't notice, but it was very recognizable, and a journalist in court recognized him.

That evening, it was all over the London papers. It drove him crazy. He was so embarrassed, so upset, that he didn't want to stay in England at all. But one interesting thing happened before he left. He belonged to the circle of Princess Margaret, who was the head of swinging London. She called and invited him to dinner.

"Ma'am, I do not want to embarrass you. I am already very ashamed about what happened."

She said, "This is not a private invitation, John, this is a royal command! A dinner in your honor. It's in the Court Circular in the morning *Times*."

Another victim of the crackdown was Sir John Gielgud—same story: Chelsea Embankment, caught and arrested, pleaded guilty, said he was a clerk, noted by a journalist, and again, front-page news. The next morning, he was in such a state, he felt he never wanted to go out in public again.

Dame Edith Evans, his star in *The Chalk Garden,* went and literally bullied him into the theatre. At the end of the play, she went offstage, took John by the hand, and walked him into center stage. The whole theatre got up and cheered and yelled. For ten minutes. It was very, very moving. □

Peter Wright is co-director of Terrain Gallery in San Francisco. These anecdotes are from a memoir-in-progress, as told to Alan Gullette, a poet who lives in Berkeley. E-mail: peterwright@telocity.com

UNDER THE STEINBECK OAK

Tad Wojnicki

I drive Main Street behind the National Steinbeck Center, watching the zooming and vrooming. Since the Center opened in 1998, I've seen visitors from all over the world, showing all possible driving styles and an infinite variety of street-crossing habits. I keep going two short blocks until I reach a seedy parking lot behind the Cherry Bean Café.

I park my '80 Mustang, grab my scratchbook, and slam the door. My car sinks well into the parking lot, sponging and fixing the smells of the neighborhood. Sucking in the air, a blend of blooming flowers and bath salts, I look down Main Street to Mount Toro. Farmland is felt. Fresh furrows get plowed nearby. I smell the sweet smell of dirt. I scent new fruit, the perfume of the globing, juicing apples, oranges, and lemons. It's so thick, I get a slight headache.

The Steinbeck Center is full-blown Postmodern, but it sits well at the end of Oldtown, lifting its dome off the globing fruit, mixing its hues with the furrows, sucking its warmth from the sunbaked Toro breast.

I like to sit at the Cherry Bean, at the bandstand table overlooking Main Street, nursing my cup and giving things a thought. For years, when I was rustling up pennies for their Steinbeck Blend, they let me steal refills. One perfumed morning, they gave me hell.

Damn good brainstorms come upon me here, like that day, three years into writing my novel, when it hit me that what I thought was my story was, in fact, not my story at all, but rather an old Bible story, "The Expulsion from Paradise." I just happened to have lived it.

Shtetl

Oldtown is about a century old, which is stinky old, if not dinosaurial, by U.S. standards. It used to be bummy and reeky, beery and pissy, but it was there, day or night. At times, I would go there at night when everything was shut. I got my kicks from the beatnik chic, which somehow zoomed me, on the gut level, back to a Polish ghetto.

Cherry Bean was fun, too, with its flower-stem lampposts in front, copied from the Marais district of Paris. Some jokers called it "Cheery Been," since it supposedly was full of cheery has-beens. Sitting there, I stewed in a Babel of tongues, steeped and brewed in the coffees and beers, and pickled in the sweat of the steerage. The stewing, brewing, and pickling healed the unhealed wounds.

Now, of course, especially since the Steinbeck Center opened, Oldtown has been scrubbed and prettified. Trendy eateries, chic boutiques. Mexican peppers, Thai spices, and sushi bars. Sidewalk bands. The banks have turned into antique malls, office buildings into art galleries. Only the Crystal Palace movie theater remains a gloomy slum.

But the old juices are still there: Sidney still dwells in his phone booth, Cherry Bean Café still pickles cheery-beens, and cat alleys still reek of undigested beer. There's even a shul down Riker Street.

If the nudnik in me takes over, I get scared. "You vant I should fix dat?" I hear God, thundering. "Lemme throw in some Cossacks for ya."

Year of the Peach

This spring adds the peach to my back-yard years. The tree has just gone blooming—overnight. Still black last night, this morning it flames pink.

The clock says 6:13 A.M. It would be an hour before my students rush in. The campus is still in the hands of the janitors. I flip *The Rite of Spring* tape over. In its spring, the symphony sounds the fall. The opposite to the expected is never expected. Smoking mug of tea in hand, I belly up to the window. Screeching jays crush the blooms, dumping the pink into the pool made by a defective sprinkler.

Each year, I rent a room, I lose a room, and I rent a room again, and each overlooks a back yard, and each back yard has a tree.

I envy them. So far, I haven't found a single spot where I could stop and whisper, "Here, Teddy, is where you'll strike a root."

Tidy as a beech or messy as a eucalyptus, big as an oak or small as a bottle-brush, all trees point to God, and, come spring, all go ablaze.

Last year, a cherry tree. I lived downtown then. Mug in hand, I stood in my back-yard window, blocked by a church painted the red of dried blood, and watched the cherry tree dripping wet and dropping petals—a jilted bride.

The year before, I jogged in the dark, checking the blooming plum before breakfast, and then I saw the klutz across the street cut it down because, he said, it grew "wild." That hurt. Tree down, me down.

When I still lived with my girlfriend, Sweets, my back yard tree was the juniper I never saw bloom. There was also a tiny fig in that back yard—a thank-you gift from my students—too young to fruit. And before that, in Carmel, a pine tree with an aluminum boat propped against it at Torres Street and Third Avenue; an apple tree bearing only rotten apples on Rio Road; a fig tree at Carpenter and First Avenue. And before that? I can't remember— it must have been some tree that year, too. This year, I go peach.

Criminal Past

I was born a criminal. The place was Poland, the year 1944. It was illegal to be born. My parents were not innocent, either. It was illegal to be alive, too. By staying alive, they turned criminals. So even before I was born, they were hiding from the law.

In 1944, both my parents were young and beautiful, just turned 20. My papa's name was Stefan, and my mother called him Stefek. Her name was Zofia, and he called her a number of endearments, including Zosha and Shurka. Though nominally Catholic, they lived in sin: my papa carrying fiery, prophetic airs about him, considered particularly "bad," made still worse by his *besserwisser,* or smart-ass, accent, and my mama by her candle-lighting (in the cupboard, of all places). Enough sinning to get burned out. Living as they lived was lawless.

Their town, Jaslo, had became too dangerous, so they fled to a godforsaken village up in the mountains, working as farmhands, and pretending to be *swoi,* or our own.

I can't figure out what it was that my mother was doing to

make a living there. She was too young to do anything, but she bragged to her sons, when we were growing up, of the "mountains" of butter cups frozen on her window sills—the village's refrigerators. Maybe she was the one who churned the butter? She probably was. But it didn't matter. She was slated for slaughter, especially after giving birth, a capital offense. If she was found out, she'd be shot dead on the spot, along with my father, and our naïve keepers.

Not me, though. For a baby, a Nazi wouldn't waste a bullet. He would grab my feet, swing me around, and hit my head against a wall.

My papa was a gifted *gonif,* or thief, I was told by people—even my grandfather, his own father, who had survived the war hiding elsewhere. If papa picked up some bad habits, there was a reason for it. Village life was harsh, brutal, and unforgiving. My papa used to be an apprentice *shohet,* or slaughterer, before the war. In hiding, he became a killer of secretly raised pigs and a maker of kielbasa, both illegal activities, punishable by execution without trial. He also made moonshine vodka. So, in 1944, my papa was a hardened criminal.

Papa was not a risk-taker, I was told, but a reckless idiot. For instance, he would throw half a pig's carcass into a wheelbarrow, cover it with hay, and, dressed like a typical villager, wheel it down into town past the German guards. In his pocket, he would carry a bottle of moonshine. Barking jokes in his street-savvy German, he would slip the bottle into a guard's pocket, and wheel away, drunk with fear. That recklessness and drunkenness stuck with him for life.

Kielbasa sold well, but there had always been one complaint. "Too hot, Stefan," clients said, as if it were a crime. "Too much garlic and onion," they added, wondering if maybe he was a Jew. Being heavy on spices was considered characteristically Jewish.

"Some like it hot," he replied.

He never changed, notoriously over-garlicking and over-onioning every dish, even after the war. I held grudges against my papa from growing up with his *meshugass,* but now I realize I'm the way he was, and my daughter is the way I've been. Each time I overdo the garlic or onion while cooking, I recall the old criminal. "Some like it hot," I say under my breath.

Keeping the Home Fires Burning

The first time I was burned out was before I was even born. It was in Jaslo, a Polish town near the Czech border. I was still in my mother's womb, and my parents had fled to the mountains.

When we came back, our house was a shell. Charred walls. Roof on the floor. Dead chickens. Smells that could kill.

The second time I was burned out, it was from the new house. I came out of the fire as naked as the day I was born. Who did it? No one has ever been found guilty. Not even accused. The neighbors thought we didn't belong. They called our lot *wygoda,* or comfort, after the pub next to it. The pub had been burned to the ground, and its owner, Shraga, killed, I was told. But I felt safe. So safe, I jeered the bullies outside my new house, daring them to beat me to a bloody pulp.

One day, I sneaked into the old Jewish cemetery after school, as I often had, to imagine burning alive. That was my obsession. I liked to transcribe the *matzeva,* or tombstone, inscriptions, copying the images—palm fronds, grapevines, hands in priestly blessings. I imagined people taking their clothes off, ordered to take a shower, but instead gassed inside. Then their clothes were burned, so nothing was left.

I got so into it, I smelled fire. Real fire. Heat pulsing. Smoke stinging. Soot flaking the air.

The stench of smoldering flesh made me puke. Scared of my overactive imagination, I ran home. But there was no home there anymore—just a red-hot shell the firefighters raked for roasted roosters.

In the Salinas Valley, I've been fired from a lot of homes. Once, for being a beatnik; another time, for burning midnight oil; and yet another, because I "didn't belong." Each time, I had to figure out how to pick up the pieces. Each time, I managed to get a new life, but I hate it. Enough lives already.

The last home, where I lived with Sweets, I lost, too, even though the only fire I experienced there was the sunshine stoking her bougainvillaea.

Where the Oranges Grow

"Go to the land where the oranges grow," my mama often said, starting when I was 13. "And then, bring us over!" I took it upon myself, making her dream come true. But the harder I tried, the harder it got, just like for my papa, who had tried to become a kosher butcher but turned into a pig-killer. Survival was self-denial, I saw.

What I didn't see was that now I would keep bumping into people blaming me for surviving, wishing they knew the taste of life in the teeth of death.

The Soviet totalitarianism I came to live under sometimes wasn't any less brutal than Nazi totalitarianism. As a teen, I had a hunger for knowledge. I wanted to write, teach, research. My ideal was to be an archeologist, deciphering tombstones, doing digs in Israel, becoming another Yigael Yadin, but no college would accept me. What was the problem? I wasn't Polish enough. Maybe one day I won't be Jewish enough, I thought. Finally, the Catholic University of Lublin wanted me. It turned out to be a great place. I was free to be who I was. I was even assigned the library pass that belonged to another Jewish student, Seweryn Blumsztajn, a dissident banned from all colleges a few years prior. The future Pope taught me ethics.

After I graduated, I was accepted as a doctoral candidate by the Institute of Philosophy and Sociology, but then, suddenly, my advisor asked that I join the Communist Party. I refused. "I'll do my best you never get a Ph.D.," he said. When I showed him my dissertation, he confiscated it—my only copy, since copying machines were illegal. I wrote another dissertation, went back to the Catholic University of Lublin, and received my Ph.D. in 1977, just days before I left Poland forever.

I had gotten married, had a daughter, all the while keeping my mama's dream alive. It had taken me 20 years to make it come true, only she had meant Israel, and I headed for California. When I was finally leaving, however, it didn't matter. She was already dead.

Independence

My first Fourth of July was tragic. Returning after the fireworks show, I got the news that my papa had died in a psychiatric hospital, my wife and daughter had been barred from joining me, and the Polish police had issued a warrant for my arrest. I burned all bridges.

Pushing Petals

Before January is over, every seed in California knows to wake up, push, and bloom. The cherry trees in front of Steinbeck House start snowing the sidewalks, magnolias' fleshy buds light up like boneyard lamps, and bougainvillaea and fuchsia burst aflame. Petals yield, fitting the bees.

Srulik of Lublin

My brother helped me come to America, but I had helped him escape first. I had been applying for postdoctoral fellowships with foreign universities, and seducing pen pals all over the world.

I had always been a maven at knocking off a killer letter, and I wrote them, forcing folks to fall helplessly for each other. Like Srulik, a Sholom Aleichem hero, writing love letters for lovers, I brought folks together while getting more and more lonely myself.

One beautiful day, I received a letter from Israel. My mama had immediately forwarded it to me, hidden in a homemade *rogaleh* package. The letter was stolen. I still recall the girl's name, Dorota Jakubowicz, and I traveled to Tel Aviv twice, but I was never able to find her.

About the same time, a perfumed letter came from a 13-year-old girl from Sacramento. She was too young for me, so I wrote a wonderful letter and signed my brother's name—he was 14. They began to write themselves. After four years, she flew in to visit. A month later, she came back again—to get married, and to take him to America. Once there, he immediately filled out the paper I needed so the secret police would give me an exit visa.

I am Circumcised

Growing up in Poland, I was stripped by bullies trying to make sure I wasn't what they called *scyzorykiem chrzczony,* or christened with a pocket-knife. During the war and even under the Soviets, being circumcised was a dead giveaway. So I hadn't been. But it was always my dream. I grew bold, bold enough to turn the tide of self-denial. When my friend Rabbi Israel Kapeluschnik, a Lubavitch Hassid, asked, "Are you circumcised?" I told him the truth.

My first big break in America came when one of the schools I had written to from Poland, Hebrew Union College in Cincinnati, gave me a postdoctoral fellowship. Then I got a reporter's job on *The Polish Daily* in New York. I started hanging out with the Lubavitch Hassidim. For the first time in America, I felt totally at home. I didn't know why, since I was born to self-deniers.

I could stay in the world of black hats, keep pious, straight, well-fed, and marry into a long line of holy *tzadikim,* or I could leave, and live recklessly, savoring the taste of the holy as well as the taste of sin.

I sat in a Crown Heights *shtibl,* a basement prayer room, over a Talmud, radiating the heat of the spring day, I knew I would always miss the drunken scent of orange blooms.

Salt of the Earth

Beneath the twin peaks, down in the belly of the valley, the Spaniards found *salinas,* salt holes. I wonder where these holes actually were. An old parchment says a property was set at *el rincon de salinas y potrero viejo.* That corner of salt holes and old pasture must be here, somewhere, in the Valley, maybe under the foundations of one of the Oldtown buildings, but where, exactly? Does the earth still swell with salt around here, anywhere?

It does.

I think one such spot is under the wine cellar of the green-and-white Victorian at Central and Stone Streets where, in 1902, the Valley's greatest son was born. John Steinbeck's emergence was a ground swell. He had to have sprung from the bowels of the Valley. This soil still sweats salt.

Burning Steinbeck

What happened at Main and San Luis Streets in Salinas goes unsung. The Salinas Public Library used to stand there; a Bank of America does now. There, in 1939, Steinbeck's books went up in smoke.

We burned them. Couldn't stand them. Describing the Valley as the hell of have-nots drove us nuts. We grabbed the books from the Library's shelves, piled them on the sidewalk, and burned the hell out of them.

Which shows, among other things, how hard it is to recognize a genius among ourselves, within ourselves. No plaque has been put at the spot, and no burner has ever stepped forward to be honored.

The library has moved to a cheaper corner, renaming itself the Steinbeck Public Library, and putting a statue of Steinbeck in front. The bronze catches him in his characteristic slouch—dragging his feet, arms dangling by his sides, out for a smoke. But the hard times to be recognized don't seem to be over—the sculptor was apparently given the wrong face: it seems he did Joseph Conrad, not Steinbeck. Recently, an anti-tobacco nut knocked the cigarette from between Steinbeck's fingers. Now he appears to be flashing the V-sign.

Live Oak

Steinbeck's shadow sprawls over this page as it sprawls over the Salinas Valley. It's the shadow of a giant live oak. He gave us our words. He used Mack to feed us the words of the underdog; Doc to teach us the lingo of loving-kindness; Danny to express our quarrels with God. That may be why in King City or Carmel, Monterey or Pacific Grove—throughout Steinbeck Country—it's a hell of a job to put two words together right.

I read about Constantin Brancusi leaving his Romanian village and crossing Europe on foot to become a student of Auguste Rodin. I think of myself, leaving my Polish village and crossing the Atlantic to become a student of John Steinbeck. But after a month of study under Rodin, Brancusi quit. "Nothing grows under a big tree," he said.

I sit at the Cherry Bean Café, a few blocks from where Steinbeck was born, and a few steps from where his books went up in smoke. My hand chases his shadow. I jot down bursts, bits, and flashes. That's what my writing is. Whatever sweetness I once had has been spoiled by salt. But it still grows. It swells. I don't care if it's bittersweet. I care that it grows—under a giant live oak, where nothing should.

Not Only, But Also

One of my high school friends, combing his hair a la James Dean, kept saying, "Steinbeck is my god." One day, I picked up *The Grapes of Wrath* in translation. I read, "Ma Joad's hands lay on her lap like tired lovers," and it knocked my socks off.

After New York, I went back to Berkeley. I took jobs as a dishwasher, waiter, orange picker, crab-meat shaker, pizza baker, whatever. I kept bumping into the great Polish poet Czeslaw

Milosz at the post office, and I began to help his T.A.s with the Polish texts. I took classes at community colleges—Vista, Laney, Cabrillo, Hartnell, Monterey Peninsula—and at the Berkeley Theological Institute, University of Texas at El Paso, University of California in Santa Cruz, and other colleges—13 in all, if my math is right.

Finally I got an M.A. at San Francisco State. Then an employment agency told me about the Defense Language Institute in Monterey. I taught there 7 years, and lately, at Hartnell College in Salinas 13 years.

Only after I arrived in the Salinas Valley did I realize Ma Joad had lived here. I sucked in the air. It smelled of lovemaking.

Smoked Out

"Did I turn off the stove after I fried my over-easy?" I think while driving to campus. I imagine the stove glowing red in the corner, setting my books afire, and reducing my complex to ashes. The thought is torture. I look up to the hills hugging the valley, their slopes bulging golden, their oaks standing still, and I wish them to become me. In one back yard, girls are chilling out. Peace overwhelms me.

I turn the radio on. "Oak trees killed by mysterious fungus add to wildfire worries," a guy says. "The dead trees stand like torches ready to turn the Salinas Valley into an inferno."

Apparently, entire neighborhoods may have to be evacuated. My apartment complex is not mentioned, but, driving to campus, I feel burned out already.

I grab the doorknob and twist it, barging in. The stove is black. The damn thing I imagined being red-hot is black—black and cold. I could kick myself for being a fruitcake, but I won't. I fold my arms and look out at the grapevines.

Love Bake

I recall my first apple tree, *bojml* in Yiddish, or *papierowka* in Polish, right in front of my boyhood window, and the beating the tree took from us boys—climbing its trunk, breaking its branches, whacking its leaves, and smacking it with hazel sticks—eager to sink our teeth in the hardly ripened apples. Every fallen fruit, sweetest of all, we crunched with the worms wiggling out. *"Dejta bojml spokoj, szkuty!"* my mother would yell from the kitchen door,

"Give the tree peace, guttersnipes!"

If I only don't get fired from my present room, I'm going to strike roots this time. I may even bake a pie, finally, a peach pie, using the peaches ripening in the back yard, the first time in my life, just for the heck of it. "It's easy," my friend says. "Just add sugar, cinnamon, and good friends. Dish all night."

Thirst

"Sal si puedes," the early Californians loved to wisecrack, describing places, "Run, if you can." After I followed the San Antonio River for what seemed like forever, and saw the creekbed, I knew it could only be Salsipuedes Creek, which meant thirst. I needed water. But there was no running away.

According to some accounts, Vincente Avila, the Mission Indian who had named the creek, reached it with his cattle, looking for pasture. Finding no water, he drove his herd up the creekbed, hoping for grass. He stumbled upon a high valley inhabited by a lone man, another Mission Indian. Desperate to save his herd, Avila offered to buy the valley, just until he could run away. He would pay all the money he had. The owner agreed. Avila shook 13 dollars out of his pockets.

"Sal si puedes," he told himself, settling down on the mesa. But he never did. He sank into that valley without water. Now, I myself am at the place of a thirst never to be quenched.

Gathering Together

My daughter is coming to visit, her third visit, next spring. She is 33, it's hard to believe, the same age I was when leaving Poland. She works as a market researcher in Warsaw. Maybe this time she will stay in America, even though she and I are as different as apples and oranges.

When she first visited, outside my window was the pine tree with the aluminum boat; the second time, the giant, gnarled fig tree. To keep her happy at Carmel Beach, by the Fisherman's Wharf, on Cannery Row was a piece of cake. But what will she say now that I've settled in a small town? Gone back into a valley just like the one I came from?

Dawn

Sleeping sweetly to the onslaught of El Niño, I hate to get up to turn off what sounds like a leaky faucet. Finally, it drives me nuts, and I get up, splashing into the water ankle-deep. The floor has been flooded. I phone the office, but get, instead of the manager, some disaster-management guy. "The Salinas River suddenly flooded at dawn," he says.

The parking lot has turned into a rapids, scrubbing the concrete clean. Junk drifts down the curbs. Leaves and needles clog the gutters. A chain saw is roaring outside my window. One of my neighbors and his buddy are hacking a fallen Monterey pine into five winters' worth of firewood. People yell from the windows. Some straddle the roaring gutters, dragging their lives wrapped in sheets. Fire trucks, their lights flashing, block the lot. Sirens bleat.

Firefighters hurry around in their yellow coats, helping Mrs. Rios and her crying kids. An ear-splitting thunderclap rattles the crowd. We're out of here—everybody, the whole damn complex.

Koby shows up, the downpour rinsing his infected nose-ring. He says his dad "swam" a mile to print a special edition of the local newspaper on the hand press on the floor the flood hadn't reached.

I yank the hard disk out of my computer; I leave everything else, and walk out, naked as the day I was born.

Somewhere, I don't know where, but somewhere here in the Salinas Valley, there is a new tree, a new back yard window, and a new room ready to ring with *The Rite of Spring*. □

Tad Wojnicki lives in Paraiso Springs and teaches at Hartnell College. His novel, Lie Under the Fig Trees, *was published by Edward R. Smallwood, Publishers, Tucson, AZ. E-mail: wojnicki@aol.com*

THE MONKEY LOOK

F. X. Toole

I stop blood. I stop it between rounds for fighters so they can stay in the fight. Blood ruins some boys. It was that way with Sonny Liston, God rest his soul. Bad as he was, he'd see his own blood and fall apart.

I'm not the one who decides when to stop the fight, and I don't stitch up cuts once the fight's over. And it's not my job to hospitalize a boy for brain damage. My job is to stop blood so the fighter can see enough to keep on fighting. I do that, maybe I save a boy's title. I do that one little thing and I'm worth every cent they pay me. I stop the blood and save the fight, the boy loves me more than he loves his daddy.

But you can't always stop it. Fight guys know this. If the cut's too deep or wide, or maybe you got a severed vein down in there, the blood keeps coming. Sometimes it takes two or three rounds to stop the blood, maybe more—the boy's heart is pumping so hard, or he cuts more. But once you get the coagulant in, sometimes you need another whack right on the cut itself. That can drive the blood away from the area, so now the stuff you're using can start to work. What I'm saying is there are all kinds of combinations down in the different layers of meat.

Some fighters cut all the time, others hardly ever. It's not something a guy can do anything about, being a bleeder, any more than a guy with a glass jaw can do something about not having a set of whiskers. I don't know if it's the bone structure around the eyes, or something to do with the thickness of the skin. Some guys get cut damn near every fight, and it doesn't take long for a bleeder's eyes to droop from severed nerves. They develop a monkey look around the eyes. Nature builds up scar tissue to protect the eyes, but in boxing the scar tissue can be the problem—the soft skin next to the scar will tear free, because of the difference in texture.

Boy gets cut, I always crack the seal of a new one-ounce bottle of adrenaline-chloride solution 1/1000. When it's fresh, it's clear like water, but with a strong chemical smell. The outdated stuff turns a light pinkish color, or a pale piss-yellow. When that happens, it couldn't stop fly blood. I might pour adrenaline into a small plastic squeeze bottle if I need to use sterile gauze pads along with a swab, but I never use adrenaline from a previous fight. I dump it, even if three quarters of it is left. This way it can't carry blood over from one fight to another, and none of my boys can get AIDS from contaminated coagulant. I'd give AIDS to myself before I'd give it to one of my boys.

I used to train fighters. But I got too old. I was walking around with my back and neck crippled up all the time from catching punches. My first fight working the corner of Hoolie Garza came after his trainer talked to me, Ike Goody. Ike was a club fighter in the fifties, but like most first-rate trainers, he was never a champ. With the exception of Floyd Patterson, who trained his adopted son, Tracy Harris Patterson, I don't remember another champ who ever trained a champion. Hoolie Garza is a smart featherweight Mexican boy who thinks he's smarter than he is. He was born in Guaymas, raised illegal in East Los Angeles. He fought with his big brothers for food. His real name is Julio César Garza, but as a kid he was nicknamed Juli—in Spanish it's pronounced *hoolie*.

After the Korean War, I went to school in Mexico City on the G.I. Bill. I wanted to learn Spanish, maybe to teach it. So I hung around with Mexicans, not Americans. Some of my friends were bullfighters. I had a fling with the daughter of the secretary to the President of Mexico, a natural blonde who drove a car with license-plate number 32. She, God bless her, was one of the ways I learned Spanish on several levels and in different accents. I usually keep my Spanish to myself, like a lot of Latinos in the U.S. keep their English to themselves. But if they find out and ask about it, I tell them I was a student in Mexico and Spain both, and I say, *Hablo el español sólo si me conviene*—I speak Spanish only when it's to my advantage. They always smile. Some laugh out loud and wag their finger.

A lot of Latino fighters coming to fight in L.A. use me in their corner; some fly me to Vegas. I'm as loyal to them as I am to an American, or to an Irishman, which is why I never bet on a fight I'm working—not on the boy I'm working with and not on the other fighter, either. This way, if I somehow screw up and cause my boy to lose, it can never be said that I did business.

Ike caught up with me at Bill Slayton's gym in South Central. "Hoolie's got a fight in Tijuana. He wants you."

"What's he getting?"

"Short money. You know about his California suspension problem? The Mexicans know about it, too. A lousy $2,500 for ten rounds. It's with a tough TJ boy, Chango Pedroza. They want to make a name off us. It's Hoolie's third fight after his suspension. Two wins by kayo. Hoolie says he'll pay the regular 2%. I told him no good, you won't work ten rounds for that, but he kept after me, so I said I'd talk to you."

"He smoking dope again?"

Ike shrugged his shoulders. "I know he's hurting for bread."

"I don't work that cheap, fifty dollars. Tell him to get someone from down there."

"He's a bleeder. That's why he wants you."

"It's 150 miles down there, Ike, so I go for a tank of gas, right? Now I don't get home until after four a.m. I don't work for fifty here in L.A., unless it's a four-rounder."

See, Ike's always told me the truth, always done square business with me, so I believe that Ike is telling me the truth about what Hoolie told him about the purse, but I know some things about Hoolie, and who's to know what kind of truth he's telling Ike? Let me tell you, Hoolie's a hell of a fighter, a tough little bastard who will meet you in the middle of the river and fight you. He's got an underslung jaw and a hooked nose that points off at an angle. And scar tissue. At 29, he's losing his hair, so he shaves his head. Tattoos from jail and from every country he's fought in, roses and daggers, same old shit. Fought for a title his third fight out of the joint, where he did time for assault with a deadly weapon. Not his hands, he didn't want to hurt his hands; he pistol-whipped some guy who smiled at his wife. He almost won his title shot, but he got tired late, and the other guy came on in the 12th. Hoolie, like always, was cut up, but the cuts didn't become a factor. After the title fight was over, Hoolie failed his piss test. They found traces of marijuana and suspended him in California for a year, and held up his purse as well. It means Hoolie can't fight anywhere else in the States that counts, because most state boxing commissions honor each other's ban. But Hoolie's a good draw, promoters from all over want him, because he's so tough and because of the blood.

That's why Hoolie has to fight for short money in Australia, in Latin America, in the Philippines, wherever there are little guys. And to stay busy, so he can be ready for his next shot at a belt.

So after Ike makes three phone calls, I settle for a hundred. I take it because Ike is a long-time friend, and because it gives me an excuse to go down to a seafood restaurant there in TJ named La Costa, a place I can get some of the best *camarrones rancheros* in the world—shrimps in hot sauce with garlic and peppers and onions and tomatoes and cilantro. Wash it down with a couple of Bohemias. For appetizers, they serve deep-fried freshwater smelt with fresh salsa and limes. I say an Act of Contrition every time I leave the place. Been going to La Costa 30 years.

I also take the fight because once the suspension is lifted, Hoolie's sure to get another title fight. He uses me, I can make a little money. Ballpark, I get first cut of the purse, two percent. Some guys get more, some less. It's business. On a $50,000 fight, that means a thousand for me. But maybe my boy doesn't get cut at all, so I just sit ringside and watch. But I still get paid. Bigger fights, I try to get the same 2% if I can, or I charge a flat fee. But a four-round prelim boy, he needs a cutman same as a champ, right? So if I'm going to be at the arena with another boy anyway, and I like the prelim boy and his trainer, or maybe I feel sorry for a scared kid, a lot of times I don't charge—the prelim boy's only making $400 in the first place. Out of that, he's got to pay his trainer 10% off the top, and his manager another 33 1/3. Ike doesn't charge his prelim boys.

But this is a game of money, right? So I got to be careful. I charge too little at the start, some boys won't respect me, and then they don't want to pay more when they make more. And some will stiff you, even after you save their careers.

Before I left Ike at Slayton's, I told him that the Tijuana Commission would look for any way to disqualify Hoolie, and to warn him that they're sure to make him take a piss test if he wins.

"You right, you right," said Ike. "Damn."

"Is he clean?"

"Say he is."

The weigh-in is at noon the day of the fight. Hoolie's staying in the same hotel where the fight's going off. He wants to eat at five, but not in the hotel, where at lunch he was pestered by people after his autograph. He's a big man in Mexico, what with him being born down there and making it in the States. He asks me about seafood and if I know a good place to eat in town. I tout him on La Costa, but tell him it isn't cheap. In TJ, he's got his wife, his mother, and two brothers he's got to feed; he's got to feed Ike and me; and Ike's back-up cornerman. There are two more to feed, a homeboy member of Hoolie's Toonerville gang and a black kick-boxer, a kid called Tweety, who's as polite and well-spoken as a Jesuit. With so many eating, it has to cost Hoolie a bundle. I wondered why he's paying for people who aren't family or working his corner, but he paid the tab without a bitch. No problem, until the waiter collected and counted Hoolie's money. I could tell from the waiter's face that Hoolie had stiffed him. So now I got to wonder if he'll do the same to me. I slip the waiter $30 for himself. With the tank of gas I had to buy, I'm working for nothing, right? The adrenaline I know I'll be using on Hoolie's cuts later that night has already cost me another fourteen dollars and change. But what am I going to do? I know these waiters for years and I can't let them get stiffed on my call.

In the second round Hoolie's eyes started to bleed. I kept him going, and as long as Ike and I could get him ready for the next round, he was standing up at the ten-second warning and waiting for the bell. Little shit, he recuperates between rounds better than anyone I ever saw. Punch by punch, he wore Pedroza down. Pedroza went after Hoolie's eyes, twisting his fists on impact to tear open the cuts even more. Hoolie stayed close, went to the body with shots to the liver, ribs, and heart. The liver shots made Pedroza gasp, the heart shots made him wobble.

Pedroza was a local boy, a good fighter with the will to win. The crowd was clearly in his corner, and so was the ref, who took a point away from Hoolie by calling a phony low blow.

In Mexico, if somebody's cut, they tend to let the fights go longer than in the U.S. But if you happen to be the guy from out of town—and you're the one who's cut—and if the promoter is looking to get a win for his boy—you know you better knock him out in a hurry, because they'll stop the fight on you as soon as they figure the local boy's ahead on points. The ref kept calling time and looking at Hoolie's cuts, but I had stopped the blood and the ref had to let him go on.

I repaired Hoolie's eyes after the third and the fourth. After the fifth, I did it again, then swabbed his nose with adrenaline to jack some energy into him through the mucous membrane. Hoolie punched himself on each side of his face and slid out to the center of the ring, his hands intentionally down low. Before Pedroza could get off on what he thought was an opening, Hoolie caught him with a sneak right-hand lead. Then he caught him with a short left hook to the liver. An uppercut put Pedroza down on the canvas. He twisted into a tight ball of hurt. The time keeper and the ref stretched the count, but they could have counted to 50 for all it mattered.

The crowd was howling and throwing beer bottles into the ring. We got to the dressing room as fast as we could. All of Hoolie's people crowded in, while Ike and I were pumping fluids into him and trying to towel him down. We were all happy and toothy. It's always like that when you win. A bottle of tequila was passed around and Hoolie took a couple of hits.

Tweety went into the crapper, turned off the light, and hid behind the partly closed door.

Two minutes later, the Commission doctor arrived, followed by the promoter whose boy Hoolie had just dropped. With a smug look, the doctor held up a plastic specimen bottle. Ike glanced over at me, rolled his eyes.

"La-la-la," said the doctor, sure he had busted Hoolie.

If Hoolie fails the test, the promoter's boy doesn't suffer the loss on his record, and the promoter doesn't have to pay Hoolie. Hoolie doesn't get paid, neither does Ike, neither do I.

Hoolie took the bottle with a smile. He went into the crapper, pushing the door ahead of him. He dropped his trunks and cup to his knees, and stood where the doctor could still see his bare ass. From my position, I could see the action. Hoolie handed the bottle to Tweety, who already had his dick out. Tweety pissed into the bottle. Hoolie sighed a piss sigh and jerked his arm around like he was shaking his dick. Hoolie took the bottle back from Tweety and handed it to the doctor.

From Hoolie's relaxed attitude, and from the heat of the specimen bottle, the doctor was no longer so sure he'd nailed an offender. The promoter saw the doctor's face, and began talking to himself.

What the doctor and the promoter were trying to do disgusted me, but the game Hoolie and Tweety were running got to me. I love boxing like I love the sacraments. You play by the rules. You never throw a fight, and you never throw intentional low blows—

unless the other guy does it first. When I realized that Hoolie was still smoking dope, I got out of there as soon as I could.

"Hoolie," I said, "I got to go. How about takin care of me."

"I'm broke until the promoter pays me, man."

"When's that?"

"Tomorrow morning when the bank opens, homes. Hey, I'm good for it, you know me, man. I don't see you around, I'll give your piece to Ike so he can take care of you, what you say?"

"It's only a hundred."

"I'm broke, man, that's why I took this shit fight, and my wife's knocked up."

I took off. I saved a doper's ass, and it cost me money. I knew then I'd never get my hundred. It wasn't enough to shoot him for, so I let it go.

It was one a.m. when I got back to the border. There were long lines waiting to get across. Vendors selling hats and serapes and pottery stood along the Mexican side. Groups of ten-year-old boys begging for change flowed like alley cats along the lines of cars; haggard women with scrawny kids sat by the roadside with their hands out. A stunted three-year-old boy stood rigidly between two lines of traffic. Tears streaked his dusty little face, snot ran down over his lips. He wailed a senseless little song and beat two small pieces of scrap wood together. Sanity had left his blue eyes. On the way home I stopped at a Denny's for coffee and a piece of gummy lemon pie.

My brother died suddenly and left me some income property on Bull Shoals Lake down on the Missouri-Arkansas border. I moved back there to fix it up and sell it. Three months after I'm in Missouri, Hoolie gives me a call. He says he's got a new trainer and a manager from Mexico. The manager's positioned him into a WBC title fight with Big Willie Little in Kansas City, Missouri.

"I want you in my corner, homes."

"Why Kansas City?"

"Big Willie's from there. It's a big deal on one of the riverboat casinos, Pay TV, all the shit."

"Why me?"

"The promoter only came up with four plane tickets, and I'm using one for my wife. That leaves tickets for my trainer and one more cornerman from out here. Besides, I don't want to chance it with some hillbilly white-bread mayonnaise sandwich from back there, right?"

"Like I say, why me?"

"You're the best, man, look what you done for me in TJ, man, they'da stopped it except for you. Besides, you're already back there, homey."

"How'd you get my number?"

"From Ike."

When I heard that Ike had given him my number, I knew Ike was scheming on the punk, that Ike wanted my presence in Kansas City, and I got interested.

"You owe me a hundred dollars, forget the gas and what else it cost me in TJ."

"I know I do, man, but you gotta know how broke I been since the suspension. It's over now, but my old lady's got cancer in the tit, *ese*, and it's costing me, but I'll give you your bread, no sweat, man."

"Is Tweety going to be there?"

"No, man, I'm squeaky clean for this one."

"Here's my deal," I said. "It's something like three hundred miles from here to Kansas City. That's all day both ways and three tanks of gas. So if I do come, I don't want to waste my time, understand?"

"No doubt about it."

"How much you gettin? Level with me."

"Yeah, yeah, only fifty grand, see? I'm takin it cheap just to get a shot at that *maiate* Big Willie mothafuck." Maiate is a word some Mexicans use for black people. A maiate is a black bug that lives in dung. "I'll take his black ass easy."

I don't trust Hoolie the fight's only for fifty thousand, not with his name on the card, but if I can make a grand, it'll buy the paint I need to finish the work on my brother's buildings.

"I'll come," I say. "But up front you send me the hundred you owe me by overnight mail. I don't get it overnight, forget it. Once I drive up to Kansas City, the day I get there you pay me a thousand up front, which is 2%. Or I turn around and come back home."

"You got it, ese, no problem, man."

"When's the fight?"

"A week from Saturday. We're flying in day after tomorrow."

"When you want me there?"

"Promoter says two days before the fight, to get your license, and all. I already got a room in your name. Your meal tickets will be at the desk."

"I don't want to lay around that long, so I'll be there one day before. Give my name to the commission at the weigh-in. I already got a Missouri license from a fight last month in St. Louis."

He gave me the name of the casino and the address. I gave him my P.O. box number and the deal was made. It took three days for my hundred to get to me, because I live way out in the hills. I cashed Hoolie's money order and drove down to Gaston's on the White River for catfish, hush puppies, and pecan pie.

The day before the fight, at six in the morning, I picked up Highway 5 out of Gainesville, and slowly headed up the climb to Mansfield. It had snowed in the night and the shivery landscape glowed in the Ozark dawn. Before the turn-off to Almartha, I watched a ten-point buck and three does race below a line of cedars, the snow kicking up like puffs of fog. Going west from Mansfield took me through the rolling hills of Amish country, black horse-drawn buggies driven along the paved shoulder by bearded men in black wearing wide-brimmed round hats. I passed through Springfield and much later on up across the backwater of the Harry S Truman Dam to Clinton.

The snow on the highway had melted because of pounding semis long before I got to a little spot called Amy Jane's Cafe in Collins, Missouri. I had two pieces of lemon pie with my coffee, which was country good. Pie and radio is how, in my family, we entertained ourselves during the great Depression. Even after World War II, when not everybody had TV sets. Picking up crumbs with my fork, I sat there thinking back. I do that more and more. I've started to miss people I've never missed before, to return to scenes from my childhood that are as fresh as if I was standing there again.

After taking the wrong exit twice in Kansas City, I got to the casino at 3:30. At the front desk they told me the weigh-in had been at noon, and that Hoolie's fight would go off at eleven the following night. From fight guys, I also learned that Big Willie Little had been three pounds overweight, had had to take them off in the steam room. Three pounds is a ton to a featherweight. It sounded good for Hoolie.

After leaving my gear off in my room, I went to the buffet, where among other things they prepared fresh Chinese food. I hadn't had good Chinese since L.A. In Springfield and Branson, and on down in Mountain Home, Arkansas, it was hog slop. The stuff in the casino was first rate and I stuffed myself. I wouldn't eat anything else that day. When I finished, I went straight up to Hoolie's room and asked for my thousand. He was playing dominoes with Policarpo Villa, a scumbag trainer from L.A. Policarpo likes to help other managers build a record for their fighters by feeding them inexperienced kids; for this he picks up a couple of hundred, a nice reward for destroying his own boys' careers. He sports a mandarin mustache that he grows down over his mouth to hide his bad teeth, and he wears a white Stetson indoors and out. It turned out that Policarpo was Hoolie's new trainer as well as his new manager. That saves Hoolie the 10% he'd have had to pay Ike, because a manager/trainer only gets 33%.

When Hoolie didn't answer me about my dough and instead kept on playing dominoes, I started tipping his pieces over so Policarpo could see his numbers.

"Hey, watchoo doin, man? I was kicking his ass!"

"We got a deal, or not?"

"I'm playin dominoes, I'm thinking, man, I got ten bucks ridin!"

"I got a grand ridin. You got my money, or not?"

"I was gonna pay you out of my training expenses, ese, but I had to pay more for sparring partners back here than I thought, you know how that goes."

"We got a deal or not?"

"We do, we do gots one. Only, look, I can only come up with three hundred now. Sparring partners back here tapped me, man, mother's honor, but you'll get the rest right after the fight when the promoter pays up, I promise."

"Do yourself a favor. Cross my name out of your chump-change address book," I said, and started for the door.

"Come on, come on, goddamnit! Don't be like that, you got to go with the flow."

Policarpo said, "Screw it. I'll be the cut man, save us both fuckin money, ese."

I laughed in his face. "You gonna handle cuts on *this* guy, and give him the right instructions in the corner in the one minute you got? You got a kit, one that's ready to go? You got all the shit? You bring adrenaline? Missouri ain't like California, you got to have a prescription for adrenaline here. And where you goin to find a drug store that even handles it? We're dealin with a bleeder, did you miss that? Go ahead, lose the fuckin fight for him, I don't give a rat's ass. I'm gonna hang around just to watch the fucker bleed."

"Calm down, calm down, ese, be cool," said Hoolie. He turned to Policarpo. "How much you got on you?"

"Two hundred, that's all I got."

Hoolie counted out his three hundred and Policarpo added two hundred more. "Here," said Hoolie. "Take it, homes, no shit, man, it's all we got until after the fight. Gimme a break, O.K.? We're gonna make big money together, you and me, word of honor."

"Gimme an IOU for the five more you owe me," I said, taking the five hundred. "You stiff me, I go to the commission."

"Hey, you write it, I sign it, that's how much I respect you, homes."

I did and he did and I left. On my way out, he asked, "When am I gonna see you?" all humble and small and best of friends. "We got to get together before the fight so I know you don't split, right?"

"You want your chiselin five hundred back?"

"I trust you, my brother, I didn't mean nothin."

"Your bout goes off at eleven. I'll be in your dressing room at nine."

"Hey, homes, no hard feelings, right?"

"Why would there be?"

The next day I slept late and took a walk down by the river. It was muddy and dark, and there were patches of foam in the weeds along the snow-covered bank. This was the river that Lewis and Clark took to open a way to the Pacific. I would love to have been along on that ride. Less than two hundred years ago, where I stood was uncharted Indian land. I wondered what kind of ride Hoolie planned for me.

I'd had a light breakfast and the cold air made me hungry. I went back for more Chinese. I was seated by the same hostess at the same table. The place wasn't crowded and I noticed for the first time that the tables were arranged in little booths made up of dividers and screens for privacy. On my way back to my table, I saw that Hoolie and Policarpo were bent over hot tea at the table next to mine. I took the long way around. They hadn't seen me, and when I sat down, I realized they were speaking Spanish. I had nothing to say to them. I'd handle the cuts, I'd collect my money, and I'd go back home and start painting. That was my deal, and I'd do it. I was kicking my own ass for showing up, but now that I was here, I was going to get my other five hundred. It was a rule.

Hungry as I was, at first I didn't pay any attention to them.

When I heard them scheming on million-dollar fights, I had to smile. Then I heard something about a two-hundred-thousand-dollar fight, and realized they were talking about the fight with Big Willie Little. I turned up both my hearing aids.

"I know they take taxes, but I don't get what we do with what's left of the two hundred thousand," said Hoolie. "The promoter said we could cash his check here if we want to, but then what? I mean, we can't pack it to L.A., right?"

Policarpo said, "Two ways. First, we could trust the promoter, and cash his check in L.A. But what if the check bounces? I say cash it here, so we got it in our hands. Then have the casino transfer the money to banks in L.A., one third to me, two thirds to you, like the big guy said."

"How much we got left over from training-expense money?" asked Hoolie.

"About thirty-five hundred. One thousand for me and two for you after the cutman gets his five."

"The cutman gets it in his ass," said Hoolie, "that's what he gets for hustling me."

"He'll be pissed, *raza.*"

"*Son cosas de la vida*—that's life."

"Can we get away with that?"

"What's the old paddy cunt gonna do?"

"You signed your name, *ese.*"

"What I signed was Julio Cercenar Bauzá, not Julio César Garza." They laughed about the one word, *cercenar*—to trim, to reduce. "Dumb old fuck didn't see the difference."

It was true. Because of Hoolie's scrawl and fancy whorls, I hadn't picked up the name switch.

"What if he says you signed it phony?" said Policarpo.

"I say I never signed it at all. He's the one who wrote the IOU, not me, right?"

"What, we just split his money, one third/two thirds?"

"No," said Hoolie, "half and half. After I kick the nig's ass, we'll go buy us some black pussy on the old man, eh?"

When they gave the high five, they saw me for the first time. I turned to one side and didn't make eye contact.

"Hey, man," said Hoolie, looking through the screen, "how long you been here?"

"Couple minutes," I said, shoveling rice into my face with chop sticks. "What's up?"

"We're gonna take a little walk, it's not too cold, and then maybe I'll have me a little siesta," said Hoolie, as he and Policarpo came around the divider. "How come you don't say hello, or nothing, man?"

"I was eatin. Didn't see you."

"Yeah, we didn't see you, too."

They stood there while I continued to eat.

Policarpo said, "You don't speak no Spanish, right?"

Hoolie's eyes flicked between Policarpo and me.

I shrugged, kept eating. "About like the rest of the California gringos," I said. "*Cerveza, puta*, and *cuánto*."

That got a laugh and they left feeling satisfied. I went back for seconds, took my time, and chewed on the fact that I should be getting four thousand dollars, not one. There were big posters of Hoolie and Big Willie in the café. More were set up throughout the hotel. This was Big Willie's fourth defense of his title and he hadn't looked good in his last one. With his weight problem, and with Hoolie's speed and boxing ability, it figured that Big Willie was due to lose. But he was a durable little battler who loved being champ. Under pressure he was mean. He would have regained his fluids since the weigh-in, and Big Willie could bang, even when he was tired. Of course, Señor Julio Cercenar Bauzá was known to bleed.

When I didn't see anyone around that was connected with the fight, I went into the casino and checked the line. Big Willie was a 3-to-1 underdog because of his weight problem. That's when I went to the nearest ATM and pulled some cash from three banks.

I looked for someone who knew me from nothing. There were hillbillies and bikers and college boys. There were sorority girls and telephone operators and welfare mothers. Old people and young. Sporting types, squares, drunks and junkies. All colors. None looked right, so I waited.

I got a whore, a skin-and-bones Thai whore with frizzed hair. She was maybe 30, but looked 50. I wondered how she could make a dime, much less pay the rent. I don't know if she was a crackhead or had AIDS, but for sure she had lived hard in the night. She made me for a typical old John, someone who wanted to feel her, not fuck her. I told her what I wanted and that I'd pay two hundred. I told her that I'd be right on her tail, that if she made a run with my money I'd stab her. She understood. What I did was slip her 15 hundred dollar bills in an envelope—to lay on Big Willie Little at the Sports Book. I win the bet, I pick up a fast forty-five hundred. Afterwards, I tailed her to a video game room. She gave me my fifteen hundred dollar print-out, and I gave her four fifties. She shoved them into her training bra.

She said, "You no wan' mo'? You no wan' bro jo'? I goo'."

I gave the poor bitch another hundred and told her to go home. She gave me a tight little smile, maybe the first she'd given in a year, maybe her last ever.

In my room, like I always do, I opened my aluminum attaché case and spread my goods out to make sure everything was there. But this time, instead of reaching for a new bottle of adrenaline, I unsnapped a flap pocket and took out an old bottle I knew had gone bad, an out-dated bottle I hadn't used from a couple of years before. It was a bottle I kept in my kit just to have a back-up bottle if I ever needed one. I'd taped the lid so I wouldn't make a mistake. When I broke the seal and poured some on a tissue, it was a pale piss-yellow. I mixed a fresh batch of salve, as I always do, using Vaseline and adrenaline. It smelled right, but the salve I prepared was from the piss-yellow stuff, not the clear. The salve's color wasn't affected. Once I made up the salve, I diluted the remaining solution with water to lighten the color. Under the ring lights, no one would notice, especially since it still smelled legit.

Even though I'm no longer a trainer, I always walk off the size of the ring. I test to see how tight or loose the ropes are. I check how hard or soft the canvas is, which is to say how fast or slow it will be. I check the steps up to the ring, how solid and wide they are, and how much room there will be at ringside. This time I checked dick.

It was a twelve-round fight and it went off on time. Hoolie and Big Willie split the first two rounds, but Hoolie came on in the third. In the fourth, each fighter knocked the other down, but neither could put the other away. Hoolie had planned to fight Big Willie from the outside, to keep him at the end of his punches, but Big Willie wouldn't cooperate. The fifth was even, but at the end of the round, Hoolie returned to the corner with a small laceration in his left eyelid. I was quick into the ring and used just enough fresh adrenaline, along with pressure, to temporarily stop the flow of blood. I also used the phony salve, which meant there would be no coagulant continually working in the wound.

Hoolie was winning the sixth easy. Near the end of the round,

Big Willie countered, whacking Hoolie on the way in with a solid one-two/one-two combination to the face, the second left-right even harder than the first. Suddenly there was a deep cut above Hoolie's right eye, and the cut in the eyelid was split wide open. The ref called time and looked at the cuts, but he let the fight continue. By the bell, Hoolie was scraping at both eyes to clear his vision.

I cleaned the wounds with sterile gauze and applied pressure with both thumbs. Once the cuts were clean, I applied some more of my out-dated piss-adrenaline.

Hoolie said, "You can fix it for me, right, homes?"

"No sweat, man."

"You're the best."

Because I had cleaned the cuts properly and because of the pressure I applied before and along with the swab, and because of the bogus salve I packed into the holes, it appeared that I had solved the problem. Policarpo and the other cornermen were so busy giving Hoolie instructions and watering him that I could have used green paint and they wouldn't have noticed.

The bell for the seventh sounded. Big Willie and Hoolie fought like bats, each turning, each twisting and bending, each moving as if suspended in light, neither stepping back, both wanting the title, both ripping mercilessly into the other. Both were splattered with Hoolie's blood. The head of each fighter was snapping back, and the ribs of both were creaking. Big Willie suffered a flash knockdown, but he was up again by the count of two. As he took the mandatory eight-count, his eyes were focused on Hoolie like a rattler's on a rat. The ref waved the fighters on. Big Willie stepped up and delivered a left-right-left combination, the second left hand snapping like it had come off a springboard. It would have destroyed most welterweights, but Hoolie grabbed Big Willie and held on.

The round ended and I cleaned the wounds and applied more pressure. I used more piss yellow.

"I thought you fixed it, ese," said Hoolie, his voice coming out small between bruised lips.

"I did fix it," I said. "But you let him pop you, so it opened up on me. Be cool. Go with the flow."

In the eighth, Big Willie looked exhausted, but there was no quit in him. He sucked it up and concentrated his shots on Hoolie's cuts. Blood filled Hoolie's eyes until he was punching blindly and getting hit no matter how he tried to cover up. People at ringside were shielding themselves from the flying blood. Big Willie saw the ruined flesh and his heart jacked up as his own adrenaline pounded through him. Walking through Hoolie's wild punches, he drilled more shots into Hoolie's blood-blind eyes. Two more cuts opened in Hoolie's eyebrows. Veins weren't cut, but blood pumped down, and the fans were yelling to the ref to stop it. He called time and waved in the ring doctor, who immediately stopped the fight.

Big Willie Little, still the featherweight champion.

In the corner, the doctor checked Hoolie's eyes. By then I had used fresh adrenaline, which stopped the blood cold. The cuts were an inch and a half, two inches long, which is big-time when it's around the eyes. But like I say, no vein was cut, and with the right stuff in there, Hoolie could have fought all night. Since Big Willie was sure to have run out of gas, and since I had no trouble stopping the cuts when I wanted to, I figured Hoolie should be the new champ. Except for me. *Son cosas de la* fucking *vida.*

Hoolie's cornermen were washing him down with alcohol and the doctor had stitched up three of the cuts when the promoter came in with Hoolie's check. He was a big round Afrikaner with a walrus mustache and a huge Dutch gut from Johannesburg. He had kind, wise eyes and seemed to float rather than walk.

"Too bad about the cuts," he said. "I thought Little was ready to go."

"I beat Big Willie's fucking ass my eyes don't go," said Hoolie, who was desolate from the loss.

"You've got one of the best cutmen I ever saw," the promoter said. "Cool under fire, he was. I watched him. Did everything right." He sucked on his mustache. "What was the grease from the little container?"

I pulled out the piss salve. I unscrewed the wide lid. "Smell."

"Ahh, yes, good lad, you mix adrenaline right into the grease, yes? Keeps working, right?"

"That's it."

"Tough break, Hoolie being a bleeder."

"Sure is. Listen," I said. "I know it's not my place, but I'm not going back to L.A. with these guys. I'm wondering if there's some way they can cash out in the casino? So they can take care of me before they take off?"

The promoter looked at Hoolie. Neither he nor Policarpo said anything.

"I've got an IOU," I said.

Hoolie saw that the promoter realized something wasn't right. He played dumb. "But once we cash the check," he asked, "we can't have the money transferred to L.A., can we?"

"Certainly can. Like I previously explained, we can arrange the transfer of funds through the casino."

"Ah, yeah, I remember now. Cool."

At the cashier's window, Policarpo counted out my money in English. "One hundred, two hundred, three hundred, four hundred, five hundred."

As he handed the bills to me, I glanced at Hoolie, whose bandaged eyes were telling me he'd never use me in his corner again. I love a guy who says he's going to fuck you because you won't let him fuck you. In his ass.

As I re-counted the first two bills in English, I decided to lay rotten eggs in Hoolie's mind. Without a break, I slipped into sing-song Mexican street-Spanish. "*Trescientos, cuatrocientos, quinientos. Correcto, mano*—three hundred, four hundred, five hundred. Correct, my brother."

Hoolie remembered our conversation over my Chinese food. "Hey, you speak Spanish?"

Now I went into a guttural, old-man Castillian. "*Pues, coño,* but only if it's to my advantage." *Pues, coño* is what nailed it—well, of course, cunt.

Hoolie blinked six times. Policarpo's jaw flopped open. For the first time I saw fear in Hoolie's eyes. Did I fuck him or didn't I?

I left him standing at attention. I showered and packed and at two in the morning went down to the casino. I saw the last of the fight guys on their way out. I pissed away a fast fifty on the quarter slots to pass time. I knew Ike had watched the fight and would know that something had gone down. We would never talk about it. I waited until three o'clock and collected my bet, plus my original fifteen. I slept for a couple of hours, had three cups of coffee in the coffee shop, and checked out.

It was 7:15 when I eased the old truck into traffic. I listened to news for a while, then switched to a jazz station that was playing Jackie McClean. I headed home the way I'd come. There was more snow on the ground, like a Christmas card.

When I got back to Collins, I pulled into Amy Jane's. Pie was in the air. A good ol' boy in a John Deere cap recognized me from the fight.

"Buddy, you looked good on TV last night. Too bad about

your boy, tough little booger."

"Real tough."

I ordered two pieces of lemon pie with my coffee, and then I found myself on the couch sitting next to my father. He was leaning into our new radio, an inlaid upright Philco with a magical green tuning light. It was June 18, 1941, at the Polo Grounds. Irish Billy Conn, the former light-heavyweight champ, and Joe Louis. Louis outweighed Conn by better than 25 pounds. In the thirteenth round, Billy went for the kill and hurt Louis early on—my father was yelling at the radio—but Louis rallied and knocked Conn out at 2:52.

At the count of ten, I watched some of my father die. As he sat with his red face in his oil-driller's hands, my mother turned off the radio. We were to eat lemon meringue pie after the fight, my father's favorite. I was able to eat a little piece, but not my da, though he tried. He fell off the wagon that night.

I finished my coffee and at the table paid the waitress.

"You didn't eat your pie."

"Lost my appetite."

I fiddled with my spoon. I sat for a while looking at my knees. I counted my keys. I fished out an El Rey Del Mundo Robusto Suprema, a hand-made maduro from Honduras that comes wrapped in white tissue. I'd fire up that spicy pup and smoke it down the highway for a good hour and a half, chew on it for more.

By the time I got up to the counter, my appetite was back. I smiled the waitress over and ordered country—a deep-fried pork tenderloin sandwich, with pickles and chips, and coffee, all to go. She didn't know *what* was going on. And pies. Two gooseberry and two rhubarb. And two lemon, too. I like tart. □

This was F. X. Toole's first time in print, in 1999, when he was 70. He had been a matador and a boxer and a cut man and a trainer, among other things. He died in 2002. His collection of stories, Rope Burns *(Ecco Press), inspired the movie* Million Dollar Baby, *directed by Clint Eastwood and starring Eastwood, Hilary Swank, and Morgan Freeman.*